SCIENCE FAIR GUIDE

This book was printed with soy-based ink on acid-free recycled content paper, containing 10% POSTCONSUMER WASTE.

HOLT, RINEHART AND WINSTON

A Harcourt Classroom Education Company

Austin • New York • Orlando • Atlanta • San Francisco • Boston • Dallas • Toronto • London

To the Teacher

The *Holt Science and Technology Science Fair Guide* has been designed to direct teachers, students, and parents through all phases of a science project. The steps outlined in this guide present a logical, detailed sequence of events which can lead to effective science projects.

The guide is organized into three parts: Resources for Teachers, Resources for Students, and Resources for Parents. Each section contains information specific to a teacher, student, or parent.

- **RESOURCES FOR TEACHERS**
  This section includes background information and tips about science projects as well as helpful planning devices, such as suggested timelines, safety information, rubrics, and progress reports.

- **RESOURCES FOR STUDENTS**
  Resources for Students is written in age-appropriate, motivational language that emphasizes fun and learning rather than competition. This section addresses every step of the process of creating a science project, from how to choose an idea to what to include in a written report. Resources for Students also includes icons that remind the student to perform certain tasks, such as writing journal entries and checking progress reports.

- **RESOURCES FOR PARENTS**
  This section provides teachers with material explaining the process to parents. Resources for Parents contains guidelines that help the parent support the child through each step of preparing his or her science project. It also outlines possible problems the parents may encounter, such as having a child who is daunted by the prospect of such a large project. Resources for Parents also encourages parents not to take possession of their child's work.

Each section walks its audience through the five phases of a science project that mirror the scientific method:

- Phase 1: Generating an Idea
- Phase 2: Research and Planning
- Phase 3: Data Collection and Analysis
- Phase 4: Writing a Report
- Phase 5: Creating and Exhibiting a Display

Copyright © by Holt, Rinehart and Winston

All rights reserved. No part of this publication may be reproduced or transmitted in any form or by any means, electronic or mechanical, including photocopy, recording, or any information storage and retrieval system, without permission in writing from the publisher.

Teachers using HOLT SCIENCE AND TECHNOLOGY may photocopy blackline masters in complete pages in sufficient quantities for classroom use only and not for resale.

Art and Photo Credits
All work, unless otherwise noted, contributed by Holt, Rinehart and Winston.
Abbreviated as follows: (t) top; (b) bottom; (c) center; (l) left; (r) right; (bkgd) background.
Front cover (owl), Kim Taylor/Bruce Coleman, Inc.; Page 37 (c), Michael Kirchhoff; 38 (b), David Merrell; 39 (c), David Merrell; 41 (b), David Merrell; 43 (b), Michael Kirchhoff; 50 (b), Michael Kirchhoff; 57 (b), David Merrell; 58 (t), Accurate Art, Inc.; 60 (t), Sam Dudgeon/HRW Photo

Printed in the United States of America

ISBN 0-03-054424-6   11 12 13   170   05 04 03

CONTENTS

Part 1: Resources for Teachers 1
Hands-on Science 2
How to Use This Guide 2
Timelines 5
    Timeline for a 4-Week Science Project 5
    Timeline for an 8-Week Science Project 6
    Create Your Own Timeline 8
Evaluation 9
    Science Project Development Rubric 12
    Science Project Presentation Rubric 13
    Create Your Own Rubric 14
    Teacher's Progress Report 15
Phase 1—Generating an Idea: Management Tips 17
Phase 2—Research and Planning: Management Tips 20
Phase 3—Data Collection and Analysis: Management Tips ... 22
Phase 4—Writing a Report: Management Tips 24
Phase 5—Creating and Exhibiting a Display: Management Tips ... 26
Additional Resources 28
SI Unit Conversion Chart 30

Part 2: Resources for Students 31
What Is a Science Project? 32
Safety Guide 33
Safety Contract 34
Student's Progress Report 35
Phase 1—Generating an Idea: Science Fair Success ... 37

SCIENCE FAIR GUIDE **iii**

CONTENTS, CONTINUED

Phase 2—Research and Planning: Science Fair Success 43

      Procedural Plan for Action 47

      Task List 48

Phase 3—Data Collection and Analysis: Science Fair Success 49

Graph Paper 54

Phase 4—Writing a Report: Science Fair Success 55

Phase 5—Creating and Exhibiting a Display: Science Fair Success . . 57

Part 3: Resources for Parents 61

Welcome to the World of Science Fairs! 62

Parent's Progress Report 65

Phase 1—Generating an Idea: Getting Involved 67

Phase 2—Research and Planning: Getting Involved 69

Phase 3—Data Collection and Analysis: Getting Involved 71

Phase 4—Writing a Report: Getting Involved 72

Phase 5—Creating and Exhibiting a Display: Getting Involved 74

Part 1: Resources for Teachers

Hands-on Science

The most obvious goal of a science project is to extend a student's understanding of science. Science is hands-on by nature, and there is no doubt that hands-on experiences facilitate the learning process. The scientific method allows scientists to collect and analyze data in a strategic and unified manner. When students use the scientific method to design and execute a project, they use the same steps professional researchers use to gather new information.

Building Scientific Confidence More importantly, science projects make science more fun and relevant to the student. Every student, regardless of aptitude, can benefit from planning and executing an inquiry-based science project. A science project can be a great way to spark a student's interest in science or to help a student develop a broader interest in research.

Philosophy of Science Fairs

Science fairs range in scale from a single class of students sharing their projects to an international event offering scholarship money as prizes. From the largest science fair to the smallest, science fairs generally share a common philosophy and similar values:

- emphasis on using the scientific method
- investigation by experimentation (inquiry-based learning)
- development of critical thinking skills
- opportunity for a positive learning experience
- extension of formal science education

How to Use This Guide

While a science project is entirely the product of a student's ingenuity and work, motivation from teachers and parents is invaluable. The *Science Fair Guide* leads teachers, parents, and students through the process of developing a science project.

What's in This Guide Using the scientific method, this guide divides the science project into five phases:

- Phase 1: Generating an Idea
- Phase 2: Research and Planning
- Phase 3: Data Collection and Analysis
- Phase 4: Writing a Report
- Phase 5: Creating and Exhibiting a Display

For each phase, you will find a set of management tips for teachers, an instruction worksheet for students, and a handout encouraging parental involvement. This guide also provides three progress reports, allowing each participant—the teacher, the student, and the parents—to monitor the student's work. Also included are sample timelines, rubrics, and safety guidelines. For additional resources on ideas and equipment, see pp. 28–29.

Rules, Rules, Rules Often a school will conduct its own science fair. The winners of the school science fair may enter a citywide science fair, and the winners of a citywide science fair may enter a regional science fair. Before assigning the project, consider whether your students should prepare their science project to meet the requirements of subsequent science fairs.

Regulations vary from science fair to science fair and even from year to year, so it is essential to contact the science fair sponsors for current listing of rules and regulations. Some science fairs have registration fees and cutoff dates, lists of materials that students are not permitted to use, and requirements for reports and displays. It is extremely important to be prepared with the necessary information before assigning a science project.

Getting Started

Before you begin Phase 1, complete these steps using the strategies that follow.

1. create a timeline
2. send home a parent letter with a safety guide, a safety contract, and a copy of the timeline
3. hand out Resources for Students and get students excited about the science fair
4. send home Resources for Parents and explain how parents should use the progress reports

1. **Timelines** Your first step is to create a timeline that will allow plenty of time for students to develop a project that is suitable for entering the chosen science fair. If students are participating in a national science fair, their projects will be more involved, requiring more teacher and parental guidance. On the next few pages, you will find sample timelines for a 4-week schedule (page 5), an 8-week schedule (pp. 6–7), and a blank timeline (page 8) that allows you to create a schedule tailored to your students' needs. Distribute this timeline to both parents and students, letting them know in advance what their responsibilities will be.

 **SCHEDULING TIP**

On the timeline, having Monday as a completion date for each assignment will allow parents to help their child over the weekend with library trips, data collection, and other parts of the project.

2. **Prepping Parents** After you have notified the students of the upcoming activity, you are ready to prepare the parents. You may choose to send them a notice 1 or 2 weeks before you make the first assignment. Include the timeline, which identifies the date

SCIENCE FAIR GUIDE **3**

of the science fair and describes when assignments are due. Also include the Safety Guide (page 33) and Safety Contract (page 34) for parents to read with their child in preparation for the experiment. Below is a sample letter.

 You can find this letter on the *One-Stop Planner CD-ROM*.

Dear Parent or Guardian,

In the next few weeks, your son or daughter will begin a science project in our science class. The science project's objective is to give every child hands-on experience using the scientific method. In addition, each student will have the opportunity to independently research a topic of his or her interest. You can help motivate your son or daughter by taking an interest in the project.

Your child will have an opportunity to enter his or her science project in the science fair, which takes place on _____. The emphasis of our project is not on winning, but on having positive learning experiences and having fun.

I am including with this letter a copy of a timeline for our class projects, a safety guide, and a safety contract. Please go over the safety guide with your child, and have him or her return the signed contract. An information packet for parents will be arriving shortly to keep you informed about how you can best assist your child with his or her project. Thank you for your help.

Sincerely,

---

3. **Motivating Students** Next, you will want to discuss with students what a science fair is and get them excited about the event. When you introduce the science fair, give the students their instruction worksheets (Part 2: Resources for Students). You can either hand out Resources for Students in its entirety or as individual worksheets as the class begins each phase. Either way, make sure students are aware of the due dates for their assignments. Also, be certain that they clearly understand the safety guidelines they are expected to uphold. Distribute the Student's Progress Reports (pp. 35–36), and explain how they will be used.

4. **Request Back-up** Finally, send home the parent information packet (Part 3 of this guide) so that the parents are prepared to help with each phase of the student's work. Be sure to explain how they can use the Parent's Progress Report (pp. 65–66) to guide their children. You will find more about how to use the progress reports in the Evaluation section of this guide (page 11).

4 HOLT SCIENCE AND TECHNOLOGY

Timeline for a 4-Week Science Project

	Sunday	Monday	Tuesday	Wednesday	Thursday	Friday	Saturday
Week project is assigned						**Phase 1** Brainstorm ideas in school and at home over weekend	
Week 1		Library research day	Develop investigative question Student-teacher meetings to confirm topic Develop hypothesis		Check progress report—end of Phase 1	**Phase 2** Library research day Reconfirm hypothesis with teacher	
Week 2			Check progress report—end of Phase 2	**Phase 3** Brief safety reminder	In-class data collection		
Week 3		In-class data collection and analysis	Check progress report—end of Phase 3 **Phase 4** Begin outline	Outline due Begin written report		In-class draft check	
Week 4		Written report due	Check progress report—end of Phase 4	**Phase 5** Reminder of display requirements	Display due Oral presentation, practice for interview. Check progress report—end of Phase 5	**Science Fair!**	

Timeline for an 8-Week Science Project

	Sunday	Monday	Tuesday	Wednesday	Thursday	Friday	Saturday
Week project is assigned						**Phase 1** Brainstorm ideas in school and at home over weekend	
Week 1		Library research day			Develop investigative question Student-teacher meetings to confirm topic Develop hypothesis		
Week 2		Check progress report—end of Phase 1	**Phase 2** Library research day	Reconfirm hypothesis with teacher			
Week 3				Library research day			
Week 4			Check progress report—end of Phase 2	**Phase 3** Brief safety reminder		In-class data collection	

6   HOLT SCIENCE AND TECHNOLOGY

Timeline for an 8-Week Science Project, continued

	Sunday	Monday	Tuesday	Wednesday	Thursday	Friday	Saturday
Week 5			In-class data collection			In-class data collection and analysis	
Week 6			Check progress report—end of Phase 3	**Phase 4** Begin outline		Outline due Begin written report	
Week 7			In-class draft check			Written report due	
Week 8		Check progress report—end of Phase 4	**Phase 5** Reminder of display requirements		Display due Oral presentation, practice for interview Check progress report—end of Phase 5	**Science Fair!**	

PART 1: RESOURCES FOR TEACHERS

SCIENCE FAIR GUIDE

Create Your Own Timeline

	Sunday	Monday	Tuesday	Wednesday	Thursday	Friday	Saturday
Week							
Week							
Week							
Week							
Week							
Week							
Week							
Week							
Week							

Evaluation

The science projects will likely be evaluated twice: first by you for a grade and second by the judges at the science fair. Setting your evaluative criteria to reflect the criteria the judges hold for a competitive project will be helpful for students who are entering the science fair. Students can then use your evaluation to improve the project before presenting it to the judges.

What the Judges Are Looking For

Because science fairs vary widely, it is of the utmost importance that you familiarize yourself with the specific judging criteria of a particular science fair. Many science fairs judge entries using the following criteria:

use of scientific thought	30%
creativity of approach	30%
thoroughness of investigation and research	15%
skill of experimental technique	15%
clarity of expression in presentation and report	10%

OTHER CONSIDERATIONS

The following are suggestions for evaluating a student project or predicting how a judge will evaluate a project:

- Besides assessing the five categories that appear above, some science fairs reward memorable presentations or displays.
- A science project should have a clear hypothesis, research plan, and conclusion.
- Many science fairs will not accept models or demonstrations, only experiments that follow the scientific method.
- Simple library research or an unplanned experiment is not acceptable.
- The student should design a controlled experiment.
- Judges typically evaluate projects using a rubric.
- Projects should be assessed against the other projects in the science fair or competition, as opposed to being judged against an ideal.
- Judges at most science fairs will be asking themselves how they would approach the investigative question and whether they would draw the same conclusions the student did.
- Creative use of materials should be considered.

Student Understanding The student's work should be evident in the project. The scientific process and the manner in which the research was conducted tend to outweigh the actual display of information. The judges will be looking carefully to see that the student has an understanding of his or her project and is responsible for the final product. In a group project, the judges will be looking for evidence that each group member completed specified tasks that furthered the progress of the group.

Tailoring Evaluation to Criteria of a Particular Fair Each science fair differs in the types of science projects that qualify for entry. Some strictly require that students perform experiments that follow the scientific method rather than perform demonstrations of scientific principles. Other science fairs accept inventions, scientific models, engineering projects, and informative exhibits. For instance, Invent America, a nonprofit K–8 educational program, holds a national contest in which students can enter inventions that solve everyday problems. The resources in this guide are tailored for experimental science projects. If you choose to broaden the scope of the projects that you accept, be sure to inform your students and their parents.

Display Regulations

Although the display may not be an important part of your evaluation, the judges will evaluate it. Rules regarding acceptable display materials vary. Most science fairs have a long list of items that are unacceptable for display. Restricted items can include everything from living organisms to any chemical, including water. It is also important to verify the parameters for displays. For example, some science fairs limit the size of displays to 76 cm deep, 122 cm wide, and 274 cm high, including the height of the table supporting the display. Be sure students have the list of unacceptable display items and the required display parameters at the time the projects are assigned.

The Poster Session Preliminary judging sometimes includes a "poster session" in which no electrical power is provided for the displays. The purpose of this session is to maintain the focus on the creative and scientific aspects of the project and to decrease the impact of a fancy display.

What to Look For

Positive Learning Many students do not feel like they can win and therefore become discouraged about participating. Emphasize healthy competition and a positive learning experience over simply winning. Participating in a science fair allows a student the opportunity to take charge of his or her own learning experience and to explore something that interests him or her. To emphasize positive learning, your evaluation can focus on the student's use of the scientific method and the amount of effort he or she put into the project rather than the student's ability to win a science fair.

Rubrics You may find that some students excel at developing a scientifically sound experiment yet are not as talented at expressing what they have learned. Others have good oral and written skills but do not have a firm grasp of the scientific method. Since both the development process and the clear expression of ideas are important aspects of a competitive science project, it may be best to focus on these skills in separate evaluations. For this reason, you will find included in this guide one rubric that emphasizes the development process and another that emphasizes the final report and display. You can use these together, or you may choose to develop your own rubric using the Create Your Own Rubric found on page 14.

 Additional checklists and rubrics can be found on the *One-Stop Planner CD-ROM*.

Progress Reports

Many students find it difficult to keep track of all of the tasks required in a project that lasts several weeks. To help you monitor student progress, you will find a Teacher's Progress Report on pp. 15–16. You may choose to use this as a checklist to determine whether or not a student performed each assigned task on time.

Student's Progress Report Maintaining the Student's Progress Report (pp. 35–36) may help your students better understand their responsibilities. The Student's Progress Report lists the tasks required in each phase of the science project, giving students a better understanding of what the judges may be looking for as well as what you expect of them. The Student's Progress Report also can serve as a handy checklist to remind students what to do next.

Parent's Progress Report A Parent's Progress Report is also included in Resources for Parents on pp. 65–66. Parents may find the progress reports useful if their child requires help with long-term organization and keeping track of responsibilities.

Progress Reports and Evaluation You may wish to include the results of all three progress reports in the final evaluation. This approach will reward students who carefully planned their science project and spent time developing each phase. Including the progress reports in your evaluation will discourage students from waiting until the weekend before the science fair to prepare the entire science project. However you choose to incorporate the progress reports into your evaluation, be sure to make both students and parents aware of the intended use of the progress reports.

Group Projects

You may choose to let your students work in groups. If this is the case, assign the groups before beginning Phase 1. For progress reports, you may still want to evaluate students on an individual basis. Making this clear at the beginning of the project may encourage participation from students who tend to let the others do the work. Consult *Holt Science & Technology Assessment Checklists & Rubrics* for a sample teacher evaluation of a cooperative group activity.

Science Project Development Rubric

This rubric focuses on the development process (planning, research, and data collection) of a science project.

Possible points	Use of Scientific Method (40 points possible)
40–31	Student designed an experiment with clear control and experimental groups that effectively tested a hypothesis.
30–21	Student designed an experiment with control and experimental groups that adequately tested a hypothesis.
20–11	Student designed an experiment with control and experimental groups that were related indirectly to the hypothesis.
10–1	Student designed an experiment that did not effectively test a hypothesis and had inadequate control and experimental groups.
	Thoroughness of Research and Data Collection (40 points possible)
40–31	Student thoroughly researched topic, collected data with care and precision, recorded all observations and sources in a science project journal, and achieved a high level of detail.
30–21	Student researched topic well, collected data with care but lack of precision, recorded most observations and sources in a science project journal, and achieved moderate detail.
20–11	Student adequately researched topic, collected data imprecisely, made a fair attempt to record observations and sources in a science project journal, and achieved average detail.
10–1	Student poorly researched topic, collected data inappropriately, and made a poor attempt to record observations and sources in a science project journal; research lacked detail.
	Originality of Approach and Use of Equipment (20 points possible)
20–16	Student adopted an original, resourceful, and novel approach and had creative design and use of equipment.
15–11	Student extended standard approach and use of equipment and exercised moderate creativity.
10–6	Student employed standard approach and use of equipment and exercised little creativity.
5–1	Student used equipment and topic in an ineffective and unimaginative way.

Science Project Presentation Rubric

This rubric focuses on the final written report and display of a science project.

Possible points	Understanding and Scientific Thought (40 points possible)
40–31	Student completely understands the topic and uses scientific terminology properly and effectively.
30–21	Student demonstrates solid understanding of the topic and adequate use of scientific terminology.
20–11	Student displays insufficient understanding of the topic and uses very little scientific terminology.
10–1	Student lacks understanding of the topic and incorrectly uses scientific terminology.

	Quality of Oral/Written Presentation (32 points possible)
32–25	Student exhibits an original, resourceful, and novel approach to presentation of topic; paper is creatively and clearly written.
24–17	Student presents topic with standard approach; writing is unimaginative but effectively gets point across.
16–9	Student's presentation of topic is incomplete and unimaginative; writing lacks clarity.
8–1	Student's presentation of topic is ineffective and lacks cohesion; paper lacks clarity and is poorly written.

	Effectiveness of Exhibit or Display (28 points possible)
28–22	Layout is logical and imaginative and can be followed easily; student displays creative use of materials.
21–15	Layout of exhibit is acceptable; student demonstrates proper use of materials.
14–8	Layout lacks organization; student exhibits poor but effective use of materials.
7–1	Layout is difficult to understand; student uses materials poorly and ineffectively.

SCIENCE FAIR GUIDE **13**

Create Your Own Rubric

Possible points	Criteria

Name _____ Date _____ Class _____

Teacher's Progress Report

❏ individual project   ❏ team member

For each step of each phase of the science project, record the task due date and the date the student accomplished the task.

Phase 1—Generating an Idea	Date due	Date accomplished
Student brainstormed five possible subjects.		
Student came up with two investigative questions for each topic.		
Student consulted with teacher and parents about project possibilities.		
Student chose a suitable topic.		
Student formed a hypothesis.		
Student discussed topic and hypothesis with teacher and gained approval.		
Student recorded ideas in science project journal.		

Phase 2—Research and Planning	Date due	Date accomplished
Student researched the hypothesis.		
Student reconfirmed or changed the hypothesis based on further research and then gained teacher approval.		
Student contacted all appropriate people before beginning data collection.		
Student recorded all details of research so far in a bibliography in the science project journal.		
Student filled out the Procedural Plan for Action and obtained necessary signatures.		
Student developed the initial plan for display materials.		

Teacher comments:

_____

_____

_____

SCIENCE FAIR GUIDE **15**

Name _____ Date _____ Class _____

Teacher's Progress Report, continued

Phase 3—Data Collection and Analysis	Date due	Date accomplished
Student conducted the experiment safely.		
Student chose an appropriate sample size.		
Student performed several trials of his or her experiment.		
Student collected data accurately.		
Student recorded all data and observations in the science project journal.		
Student graphed or charted data and looked for trends.		
Student prepared a written conclusion supported by the data.		

Phase 4—Writing a Report	Date due	Date accomplished
Student answered the questions on page 55.		
Student prepared an outline and discussed it with the teacher.		
Student prepared a draft and discussed it with the teacher.		
Student revised the draft according to the teacher's feedback.		
Student turned in the final draft of the written report.		

Phase 5—Creating and Exhibiting a Display	Date due	Date accomplished
Student sketched possible designs for display.		
Student created a display board within the appropriate parameters.		
Student displayed the results in a clear and interesting manner.		
Student gave an oral presentation as practice for the science fair interview.		

16 HOLT SCIENCE AND TECHNOLOGY

Phase 1—Generating an Idea: Management Tips

> **DURING THIS PHASE, STUDENTS WILL**
>
> 1. brainstorm five possible subjects
> 2. come up with two investigative questions per topic
> 3. consult with teacher and parents about project possibilities
> 4. choose a suitable topic and investigative question
> 5. form a hypothesis
> 6. discuss topic and hypothesis with the teacher and gain approval
> 7. record ideas in their science project journal

Begin by distributing the Phase 1 packet to students and parents. Check the timeline and schedule library visits and individual meetings with students.

> **TIP**
>
> Throughout the course of the project, and especially during the beginning stages, emphasize that one of the main goals of the project is to have fun. Healthy competition can be exciting, but the purpose of a science project is to learn about science through inquiry in a hands-on experiment.

The Journal: An Important Tool Resources for Students periodically reminds students to enter information in their science project journals. Students will need their science project journals as soon as they begin Phase 1. They may use a spiral bound notebook, or they may choose to make a journal by stapling together ruled paper or graph paper.

Explain to students that this journal is an important tool that they will use in every phase of the science project. For instance, they will write in their journal all of the notes that they take while brainstorming and researching. They will also record all data, observations, and calculations in their journals. This way all of the information needed for writing a report and preparing a display will be in one place. For group projects, every team member must keep a science project journal containing their notes and data.

SCIENCE FAIR GUIDE 17

Brainstorming for Topics

Students often find that coming up with a suitable idea is the most challenging part of a project. Encourage students to brainstorm and to be creative. Trying to come up with new ideas can be difficult and sometimes embarrassing for sensitive students. Make sure they feel free to record any idea, no matter how odd it seems at first. Familiarize yourself with the students' packets to help them. If you know of a subject that a student is particularly interested in, offer ideas related to that topic.

Caution: *Be careful not to take over the brainstorming process for students.* Avoid assigning topics. Unless a student finds something that interests him or her, he or she may have a difficult time putting forth the effort necessary to completing a project.

Suggesting Ideas For students who have trouble coming up with science project topics, try the following:

- Provide your students with a list of projects that previous students have successfully completed.

- You may also decide to list some projects that did not work out as planned. Explain why the projects were unsuccessful (too involved, too expensive, not enough time, etc.).

- *Holt Science & Technology Long-Term Projects & Research Ideas* contains several ideas that may be developed into science projects.

- Resources for Students mentions several topics for science projects.

- The Additional Resources section of this booklet also provides a list of books with ideas for science projects (see pp. 28–29).

Forming Investigative Questions

Require that all students generate at least two investigative questions for each of five topics. If they have 10 investigative questions to choose from, chances are good that they will be able to settle on a viable investigative question.

 **TIP**

> The timelines and Resources for Students suggest a library trip early in Phase 1, after a student has brainstormed for topics. Before or during the scheduled library trips, brief students on how to use library resources.

Strengthening Research Skills For students who require additional help with researching topics, use the following worksheets from *Holt Science & Technology Science Skills Worksheets:*

- Worksheet 17: Organizing Your Research
- Worksheet 18: Finding Useful Sources
- Worksheet 19: Researching on the Web

The Teacher-Student Meetings

This is the best time to meet with students individually to ensure that they are on the right track. Discuss their project options, and help them decide which investigative question they will choose. Steer them away from projects that would exceed a reasonable budget or projects that may risk disqualification. During the meetings, it would be helpful for you to have a copy of the science fair rules handy to point out problems that may make a particular project difficult. For example, many science fairs prohibit the display of chemicals. A student may be allowed to perform a supervised procedure with chemicals during the data collection phase of the project, but the rules prohibiting chemicals would limit the display.

Forming a Hypothesis

When students have chosen their topics, it is time for them to form a hypothesis. Having researched their topic thoroughly, students are now able to make an educated guess that answers their investigative question. Make sure students understand the scientific method. Emphasize that they will not design an experiment in order to yield results that *support* the hypothesis they have formed. Instead, they will design an experiment to *test* the hypothesis. There is often great pressure to be right about the hypothesis, but that is not the point of a science project. The purpose is to discover whether the hypothesis is supported or disproved by the experimental results. The quality of the project is independent of the accuracy of the hypothesis.

✓ At the end of Phase 1, make sure you are still following the timeline. Check students' progress on their Student's Progress Report and Parent's Progress Report.

Phase 2—Research and Planning: Management Tips

> **DURING THIS PHASE, STUDENTS WILL**
>
> 1. research the hypothesis
> 2. change or reconfirm the hypothesis with the teacher
> 3. contact all appropriate people
> 4. record all research sources in their science project journal
> 5. fill out the Procedural Plan for Action and obtain signatures
> 6. develop the initial plan for display materials

Research

You may feel it is necessary to give students a further lesson on Internet research or library research. The library may have a list of useful sources that students can refer to while planning their experiments. You may wish to meet with the librarian to discuss what the students will be researching so that he or she can prepare for the class visit. Also, Additional Resources (page 29) provides a list that includes books about developing science projects. Remind students to record all research in their science project journals.

Students may wish to obtain information using other methods, such as seeking an interview with a scientist or writing a letter to a company. Encourage students to contact all sources early in Phase 2 in order to allow plenty of time for a response. The student packet includes a sample letter for students who wish to contact someone about information needed for their science project.

Bibliography Make sure students are familiar with the style of bibliography that you require. Resources for Students encourages students to record all sources used, including interviews, in their science project journals. Recording all sources during the research phase will give students a head start on the bibliographies that they will need for the written report and the display.

Procedural Plan

Students will plan how they are going to test their investigative questions. Their packets include a brief discussion of control groups versus experimental groups and dependent variables versus independent variables. You may want to discuss these terms further in class. Once each student has an idea of how to scientifically test his or her hypothesis, you may require students to fill out the Procedural Plan for Action (page 47) and the Task List (page 48).

Caution: *Watch for projects that will require too much time, money, or effort.* Now is the time for the student to determine whether the procedural plan for an experiment exceeds budget or time limits. At this point, it would not be too late for a student to choose a different topic. However, changing a project any later than this stage may pose a problem in meeting the deadlines on the timeline.

Obtaining Equipment and Materials

Cutting Costs Remind students to keep cost and availability of materials in mind when planning a project. Some ideas require more resources than are available within a limited budget. Not all expensive ideas need to be discarded. Encourage students to come up with less expensive ways to conduct their experiment or ways to perform a similar yet less costly experiment within the chosen topic. For prices and availability of materials, provide students with a list of local businesses that carry scientific and laboratory supplies (see page 28).

After students have chosen a hypothesis and planned their project, it is important to anticipate costs. When preparing a budget, a student should consider the following expenses:

- Science fair entry fee
- Science fair travel expenses
- Library trips
- Book purchases
- Journal purchase
- Materials necessary for experimentation
- Materials necessary for display

Creativity Counts Judges often take creative use of materials into account when judging a science fair; a student who has built a barometer will likely be recognized as more motivated than a student who purchased one. In other words, high-dollar projects will not be recognized as better just because they employ fancier equipment. The library and Internet may provide resources with suggestions on how to make equipment using inexpensive materials.

Outside Sponsors Also, many businesses and institutions will lend equipment to young scientists. Encourage your students to write to such businesses for help in obtaining expensive equipment or materials for their science projects. Be sure that the schedule allows enough time for students to receive a response. (This approach to procuring equipment works best with an 8-week timeline.)

MONEY-RAISING IDEAS

The following are a few ideas for raising funds to help defray science project expenses:

- A local business may be willing to act as a monetary sponsor in exchange for an advertisement at the event or in the event program.
- Schools participating in science fairs may hold raffles to raise money for projects.
- Parents may be willing to donate materials (or time) to support the science fair.

✓ At the end of Phase 2, make sure you are still following the timeline. Check students' progress on their Student's Progress Report and Parent's Progress Report.

Phase 3—Data Collection and Analysis: Management Tips

> **DURING THIS PHASE, STUDENTS WILL**
>
> 1. conduct experiments safely
> 2. choose an appropriate sample size
> 3. perform several trials of the experiment
> 4. collect data accurately
> 5. record all data and observations in their science project journal
> 6. graph or chart the data and look for trends
> 7. prepare a written conclusion supported by the data

Safety First! Safety is of particular concern during the data collection phase. The timelines allow one class period for a brief review of safety before students begin collecting data. The in-class data collection days give you a chance to supervise safety. Remind students to have an adult supervise all data collection at home.

In-Class Data Collection

The following tips may be helpful during in-class data collection:

- Take a moment in class to review the difference between qualitative data and quantitative data. Go over with your students the kinds of projects that call for each type of data.

- Make sure that students understand the importance of using an adequate sample size, as discussed in Resources for Students on page 50.

- Explain that in order for data to be valid, a researcher must perform several trials of an experiment.

- Watch to see that students measure accurately. (You may choose to distribute the SI Unit Conversion Chart on page 30.)

- Check students' science project journals to make sure that they keep a neat record of all aspects of the project.

- If a student's project does not lend itself to collecting data in class, ask the student to bring in a book he or she is reading for continued research so that class time will be used effectively.

- Students who have collected all the data they need can use the in-class data collection days to work on analyzing the data.

22 HOLT SCIENCE AND TECHNOLOGY

Strengthening Data Collection Skills For students who require additional help with data collection, use the following worksheets from *Holt Science & Technology Science Skills Worksheets:*

- Worksheet 14: Using the International System of Units (SI)
- Worksheet 15: Measuring
- Worksheet 23: Science Drawing

Analysis and Conclusions

Analysis When students have finished collecting data and making observations, they can begin the analysis. As part of the analysis, students will make charts and graphs in their science project journals. Page 54 includes a sheet of graph paper that can be used for graphing purposes. Resources for Students discusses graphing, which you may want to review in class, along with a brief lesson on useful mathematical terms, such as *median* and *mean.* Instruct students to examine the charts and graphs for trends and record these trends and all calculations in their journal.

Strengthening Graphing and Analysis Skills For students who require additional help with graphing and analysis, use the following worksheets from *Holt Science & Technology Science Skills Worksheets:*

- Worksheet 25: Introduction to Graphs
- Worksheet 26: Grasping Graphing
- Worksheet 27: Interpreting Your Data
- Worksheet 28: Recognizing Bias in Graphs
- Worksheet 29: Making Data *Mean*ingful

Conclusions In the conclusion, students will determine whether the results support or disprove the hypothesis. Discourage students from simply writing opinions in the conclusion. Instead have students discuss factors that contributed to the results and explain how they would control these factors if the experiment were performed again.

✓ At the end of Phase 3, make sure you are still following the timeline. Check students' progress on their Student's Progress Report and Parent's Progress Report.

Phase 4—Writing a Report: Management Tips

> **DURING THIS PHASE, STUDENTS WILL**
> 1. answer the questions on page 55
> 2. prepare an outline and discuss it with the teacher
> 3. prepare a draft and discuss it with the teacher
> 4. revise the draft according to the teacher's feedback
> 5. turn in a completed draft

Assigning a Written Report If the science fair you are entering requires a written report, you will need to share the particulars (how long it should be, whether it must be typed, etc.) with your students and their parents. Not all science fairs require a written report, but a paper gives students the chance to express what about the project meant the most to them and to organize their thoughts for an oral presentation. It is also a very important part of any scientific endeavor.

Having Fun Have your students use the italicized questions on page 55 of Resources for Students as a guideline for writing their reports. By asking students to answer questions about their personal experiences with the emphasis on what they enjoyed about doing a science project, you can reinforce that the science fair can be fun. The more fun they have writing the report, the easier it will be to complete and the more interesting it will be to read.

Writing Tips

Use the following tips to help students with writing:

- Require students to first write an outline for the written report. You may need to review the outlining process in class.

- Explain how index cards that are grouped together by subject can be helpful in writing a paper.

- Written reports can greatly improve when students write more than one draft. You may wish to add more drafts to the timeline.

Strengthening Writing Skills For students who require additional help with science writing, use Worksheet 22: Science Writing from *Holt Science & Technology Science Skills Worksheets*.

Feedback and Assessment

Giving feedback to a student need not be difficult. The following tips may help:

- Try to mention both the strong and weak points of a draft.
- Encourage students to do several drafts, if needed, and to ask an adult to read each one.
- Peer editing may be helpful and will give students a chance to see how their friends are handling similar problems.
- When assessing a written report, be aware that students' different writing abilities can affect the overall quality of presentation. Try not to let that overshadow the quality of the scientific content.

✓ At the end of Phase 4, make sure you are still following the timeline. Check students' progress on their Student's Progress Report and Parent's Progress Report.

Phase 5—Creating and Exhibiting a Display: Management Tips

> **DURING THIS PHASE, STUDENTS WILL**
> 1. sketch possible designs for the display
> 2. create a display board within the appropriate parameters
> 3. display results in a clear and interesting manner
> 4. give an oral presentation as practice for the science fair interview

The Display Board

A typical display is a trifold board made from corrugated cardboard, foam core, or corkboard. Impress upon students the importance of measuring the display board more than once to ensure that it fits the size constraints of the science fair. Many science fairs require that the display be no more than 274 cm tall, 122 cm wide, and 76 cm deep. Having an oversized display can disqualify the entire project, no matter the quality of the work. Be aware that certain materials are often prohibited in the display area.

A Sample Display You may want to set up a successful sample display in your classroom. Students can study the sample display up close, noticing the neatness of the lettering, the layout of the display, and the type of information contained in each section. Show students how a display board can have a large middle section and two smaller "wings" on the left and right that can fold up, making a display more portable.

Is the Data Clearly Presented and Relevant?

You may decide to have conferences with students about the data that they plan to present. Check charts and graphs for accuracy and readability. Make sure that the information shown in the figures is relevant to the purpose, hypothesis, and conclusion. The purpose of a display is to present the information in the clearest manner possible so that the judges will be able to recognize quickly that the student performed a successful science project.

Display Design Although the display introduces the project, the design should not distract from the content. Encourage students to be creative with borders, font, and layout but not to the point that a judge would find it difficult to read the information contained in the display. Illustrations should be informative, not just decorative. A sleek, mature, and professional style can impress the judges, but they prefer presentations which are unique and are clearly a product of the student who created it.

Preparing for the Interview

Many science fairs require that a student give a presentation about his or her project to the judges. The presentation summarizes each step of the science project: why the student chose his or her subject, a statement of the hypothesis, what kind of data was collected, a brief summary of the data, and the conclusions that the student came to when he or she analyzed the data. Students can also discuss how they would do the experiment differently if they were to start over again or what other questions arose during their research.

A Captive Audience If the science fair your students are entering requires an interview, it is in your students' best interest to practice the presentations at least once before the science fair. Requiring an oral presentation in front of the class can be beneficial to all students. It will give them an opportunity to practice speaking in front of a group and to determine the most effective presentation styles by watching each other. An oral presentation will give you an opportunity to evaluate what students have done, as well as give personal feedback that will be very helpful during the interview with the judges.

During the interview, the judges will likely ask students a few questions. To prepare your students for this, ask questions during their oral presentation to the class. Encourage other students to ask questions as well. Also, suggest that they make a list of possible questions the judges could ask and possible answers to those questions.

Practice, and Practice Again Some students dread the interview part of the science fair. The more practice a student gets for the interview, the more comfortable he or she will be speaking to the judges. Encourage students to practice as often as possible and in front of as many people as possible. For instance, students may want to practice by giving their presentation to family members at home. Remind them to relax and have fun.

✓ At the end of Phase 5, make sure you are still following the timeline. Check students' progress on their Student's Progress Report and Parent's Progress Report.

Additional Resources

On the next few pages, you will find resources that may be helpful in the science fair process. Included are:

- a list of suppliers of scientific laboratory equipment
- the address and phone number of Science Service, a company that supports and regulates science fair administration
- a bibliography of books that can help students choose, design, and complete a science project
- a standard SI unit conversion chart to help students learn to work in the SI system

SCIENCE FAIR GUIDE **27**

Additional Resources

The following list is a compilation of resources for students and teachers.

Equipment Suppliers

The following is a list of a few scientific supply companies that specialize in laboratory equipment:

Carolina Biological Supply Co.
2600 York Road
Burlington, NC 27215
(800) 334-5551

Custom Lab Supply
801 98th Avenue
Oakland, CA 94603
(510) 633-1329

Science Kit and Boreal Labs
777 East Park Drive
Tonawanda, NY 14150
(800) 828-7777

WARD's Natural Science Establishment, Inc.
5100 W. Henrietta Rd.
Rochester, NY 14585
(800) 962-2660

Science Fair Administration

Science Service, Inc.
1719 North Street, NW
Washington, DC 20026
(202) 785-2255

Books

Beller, Joel. *So You Want to Do a Science Project!* Paramus, N.J.: Prentice Hall, 1984.

Bochinski, Julianne. *The Complete Handbook of Science Fair Projects.* New York: John Wiley & Sons, 1996.

Bombaugh, Ruth. *Science Fair Success.* Berkeley Heights, N.J.: Enslow Publishers, 1990.

Bonnet, Robert and Daniel Keen. *Botany: 49 Science Fair Projects.* New York: McGraw-Hill, 1989.

Bonnet, Robert and Daniel Keen. *Earth Science: 49 Science Fair Projects.* New York: McGraw-Hill, 1990.

Brisk, Marion. *1001 Ideas for Science Projects.* New York: Macmillan, 1992.

Cook, James. *The Thomas Edison Book of Easy and Incredible Experiments.* New York: John Wiley & Sons, 1988.

Iritz, Maxine. *Science Fair: Developing a Successful and Fun Project.* New York: McGraw-Hill, 1987.

Mandell, Muriel. *Simple Science Experiments with Everyday Materials.* New York: Sterling Publishing, 1989.

Tant, Carl. Projects: *Making Hands-On Science Easy.* Angleton, Texas: Biotech Publishing, 1992.

Tocci, Salvatore. *How to Do a Science Fair Project.* Danbury, Conn.: Franklin Watts, 1989.

Wolfe, Connie. *Search: A Research Guide for Science Fairs and Independent Study.* Tucson, Ariz.: Zephyr Press, 1988.

Wood, Robert. *Physics for Kids: 49 Easy Experiments with Heat.* Blue Ridge Summit, Penn.: TAB Books, 1990.

SI Conversion Table

SI Units	From SI to English	From English to SI
Length		
kilometer (km) = 1,000 m	1 km = 0.621 mile	1 mile = 1.609 km
meter (m) = 100 cm	1 m = 3.281 feet	1 foot = 0.305 m
centimeter (cm) = 0.01 m	1 cm = 0.394 inch	1 inch = 2.540 cm
millimeter (mm) = 0.001 m	1 mm = 0.039 inch	
micrometer (μm) = 0.000 001 m		
nanometer (nm) = 0.000 000 001 m		
Area		
square kilometer (km^2) = 100 hectares	1 km^2 = 0.386 square mile	1 square mile = 2.590 km^2
hectare (ha) = 10,000 m^2	1 ha = 2.471 acres	1 acre = 0.405 ha
square meter (m^2) = 10,000 cm^2	1 m^2 = 10.765 square feet	1 square foot = 0.093 m^2
square centimeter (cm^2) = 100 mm^2	1 cm^2 = 0.155 square inch	1 square inch = 6.452 cm^2
Volume		
liter (L) = 1,000 mL = 1 dm^3	1 L = 2.113 pints	1 pint = 0.473 L
liter (L) = 1,000 mL = 1 dm^3	1 L = 1.057 fluid quarts	1 fluid quart = 0.946 L
liter (L) = 1,000 mL = 1 dm^3	1 L = 0.264 gallons	1 gallon = 3.785 L
milliliter (mL) = 0.001 L = 1 cm^3	1 mL = 0.034 fluid ounce	1 fluid ounce = 29.575 mL
microliter (μL) = 0.000 001 L		
Mass		
kilogram (kg) = 1,000 g	1 kg = 2.205 pounds	1 pound = 0.454 kg
gram (g) = 1,000 mg	1 g = 0.035 ounce	1 ounce = 28.349 g
milligram (mg) = 0.001 g		
microgram (μg) = 0.000 001 g		
Temperature		
Freezing point of water	32°F	=	0°C
Boiling point of water	212°F	=	100°C
To convert °F to °C	(°F − 32) × 5/9	=	°C
To convert °C to °F	(°C × 9/5) + 32	=	°F

Part 2: Resources for Students

What Is a Science Project?

Scientists all over the world make new discoveries by using the scientific method. Now, you can join their ranks! A science project is your chance to choose a subject of interest to you and study it in the exact same way that a professional scientific researcher would.

You may choose to study a topic such as how lighting affects plants or how much of the air is oxygen. Whatever you choose to do, the worksheets on pp. 35–60 will lead you through each step of the scientific method and through each phase of the project. The worksheets will even help you choose your subject if you can't decide what to do. So, don't worry! The project is entirely yours. Your parents and teachers will be around to help you if you need it. You'll also have plenty of tools such as timelines and progress reports to help keep you on track.

Explore Your World The whole point of a science fair is to give you a chance to explore on your own. Exploration can be a lot of fun. Instead of reading about composting in a book, you can do experiments comparing store-bought fertilizer to compost that you have made. You may try to figure out what weather patterns would be like if the Earth were square. Or, you can discover what types of plants attract butterflies. Maybe you'll build a machine to test the difference between old golf balls and new golf balls. The fact is that your science project presents *your* creative solution to a question or problem. So good luck, and have a blast!

THE SCIENTIFIC METHOD

The steps of the scientific method that you'll be using appear below. Each step is explained in Phases 1–5 of this packet.*

Purpose: developing an investigative question

Hypothesis: making an educated guess about the answer to the investigative question based on research

Experiment: testing the hypothesis, collecting data, and making observations

Analysis: organizing data from the research and experimentation and looking for patterns

Conclusion: determining if the hypothesis is supported or disproved by the experimental results

Communicating the Results: sharing the conclusion with others

*In Chapter 1 of your textbook, these steps are called Ask a Question, Form a Hypothesis, Test the Hypothesis, Analyze the Results, Draw Conclusions, and Communicate Results.

Safety Guide

Science is a lot of fun, and you'll have the most fun if you avoid accidents. Some simple precautions can go a long way to ensure the safe and successful completion of your project.

The major causes of laboratory accidents are carelessness, lack of attention, and inappropriate behavior. Following the safety guidelines below will greatly reduce your chances of having an accident. While you are working on a science experiment at home, even a minor accident can cause serious injuries, so be very careful.

- Know the locations of the fire extinguisher, telephone, and first-aid kit in the event of an emergency.

- Always have an adult (parent or teacher) supervising the data collection phase of your science experiment.

- Wear safety goggles and tie back loose hair and clothing when working with any chemical, flame, or heating device.

- Wear an apron and gloves when using acids and bases.

- Never smell or taste a chemical unless instructed to do so by your teacher.

- Never use an electrical device with a frayed cord. Never use an electrical appliance with wet hands or with water nearby.

- Never eat any part of a plant used in an experiment.

- Whenever possible, use plastic test tubes, beakers, and flasks. Check all glassware for chips and cracks. Glass containers used for heating should be made of heat-resistant glass.

- Whenever possible, use a hot plate rather than an open flame or burner. Make sure to turn off and unplug a heating device when you are through with it.

- Check with your state board of education before experimenting with and exhibiting animals. Permits and/or veterinarian supervision may be required. Also, wash your hands with hot water and soap after touching any animal.

- Students and adults should wear ultraviolet safety goggles during operation of UV light.

- Do not use cultures from any warm-blooded animal.

- If you are planning on using X rays, cathode ray tubes, or radioactive substances, you must get information on federal guidelines for their use. Consult the Consumer Affairs Office of the Center for Devices and Radiological Health, a division of the Department of Health and Human Services.

- Many states require registration of lasers. Check with the state board of education for tips on how to register.

- Discuss the safety of your materials with your teacher or another scientist.

Safety Contract

I, _____, hereby certify that on this day of _____, I have successfully completed a review of safety procedures for a science project. I agree to follow the safety guidelines listed below, and I will take every necessary precaution to operate safely throughout my experiment.

- I will follow the safety guidelines of my teacher and my school.

- I will keep my work area neat and free of unnecessary papers, books, and materials. I will keep my clothing and hair neat and out of the way, and I will wear a safety apron and/or gloves if necessary.

- I know the location of all safety equipment (such as the fire extinguisher and first-aid kit) and the nearest telephone.

- I will wear safety goggles when handling chemicals, working with a flame, or performing any other activity that may cause harm to my eyes.

- I will not use chemicals, heat, electricity, or sharp objects until my teacher or parent instructs me to do so, and I will follow the adult's instructions carefully.

- I will be especially careful when using glassware. Before heating glassware, I will make sure that it is made of heat-resistant material, and I will never use cracked or chipped glassware.

- I will wash my hands immediately after handling hazardous materials. I will clean up all work areas before I leave the laboratory, put away all equipment and supplies, and turn off all water faucets, gas outlets, burners, and electric hot plates.

I understand and agree to the above and all other safety precautions presented to me in class. I am hereby ready to undertake my science project with safety from this day forward.

_____
Student's signature

_____   _____
Teacher's signature   *Parent's/guardian's signature*

Name _____ Date _____ Class _____

Student's Progress Report

For each step of each phase of the science project, mark the date it is due and the date you completed your work. Reward yourself for your hard work!

Phase 1—Generating an Idea	Date due	Date accomplished
I brainstormed five possible topics.		
I came up with two investigative questions for each topic.		
I consulted with my teacher and parents about project possibilities.		
I chose a suitable topic.		
I formed a hypothesis.		
I discussed topic and hypothesis with my teacher and gained approval.		
I recorded ideas in my science project journal.		

HOORAY! You've completed Phase 1—now you're on your way!

Phase 2—Researching and Planning	Date due	Date accomplished
I researched my hypothesis.		
I reconfirmed or changed my hypothesis based on further research and then gained teacher approval.		
I contacted all appropriate people before beginning data collection.		
I recorded all details of research so far in a bibliography in my science project journal.		
I filled out the Procedural Plan for Action and obtained necessary signatures.		
I developed the initial plan for display materials.		

WAY TO GO! You've completed Phase 2—give yourself a pat on the back!

Comments: _____

_____

_____

SCIENCE FAIR GUIDE

Name _____ Date _____ Class _____

Student's Progress Report, continued

Phase 3—Data Collection and Analysis	Date due	Date accomplished
I conducted the experiment safely.		
I chose an appropriate sample size.		
I performed several trials of my experiment.		
I collected data accurately.		
I recorded all data and observations in my science project journal.		
I graphed or charted data and looked for trends.		
I prepared a written conclusion supported by the data.		

ALL RIGHT! You've completed Phase 3—you're halfway there!

Phase 4—Writing a Report	Date due	Date accomplished
I answered the questions on page 55.		
I prepared an outline and discussed it with my teacher.		
I prepared a draft and discussed it with my teacher.		
I revised the draft according to my teacher's feedback.		
I turned in the final draft of my written report.		

FANTASTIC! You've completed Phase 4—you're almost done!

Phase 5—Creating and Exhibiting a Display	Date due	Date accomplished
I sketched possible designs for my display.		
I created a display board within the appropriate parameters.		
I displayed the results in a clear and interesting manner.		
I gave an oral presentation as practice for the science fair interview.		

CONGRATULATIONS! You've completed your science project!

Phase 1—Generating an Idea: Science Fair Success

> **DURING THIS PHASE, YOU WILL**
>
> 1. brainstorm five possible subjects
> 2. come up with two investigative questions per topic
> 3. consult with your teacher and parents about project possibilities
> 4. choose a suitable topic and investigative question
> 5. form a hypothesis
> 6. discuss topic and hypothesis with your teacher and gain approval
> 7. record ideas in your science project journal

Remember to update your Student's Progress Report as you go along. You will have until _____ to finish Phase 1.

Your Journal: The Most Important Tool You'll Use
The first thing you'll need to do is sit down with your science project journal and brainstorm a list of possible science project topics. Keep all notes that you take and data that you collect in this journal. The point of having a science project journal is simple: if you write everything down in one place, you'll know where all of your information is during the entire project. It will also be much easier for you to show your progress to your teacher and parents. Because all of your project information will be in your journal, be sure to keep it in a safe place.

Brainstorming for Topics

Here's where it all begins. By following the steps outlined in this section, you will take a general idea and narrow it down into a working science project. You will start with a general topic, choose a specific research question, and then develop the hypothesis that your experiment will test.

SCIENCE FAIR GUIDE **37**

Ideas Galore! Brainstorming means coming up with as many interesting ideas as possible and writing them down in a list. Don't worry about how your ideas sound or whether you will actually pursue each topic. The important thing at this stage is to come up with a lot of ideas to choose from. The more ideas you have, the more likely it is that you will find one that is interesting and one that you'll enjoy working on for the entire length of the project.

Make It Fun Since you'll be spending weeks or months preparing for the science fair, you really want to be sure that you have a true interest in finding out about your topic. Even though all of what you learn in school is worthwhile, most of the topics have been chosen by someone else. Take full advantage of the chance to choose something you're really interested in for your science project.

Getting Started

Find Out More To begin the brainstorming process, think about scientific topics you've always wanted to know more about. Maybe you've always been interested in the way mirrors work or the way light waves travel. Have you been on a field trip recently that sparked your interest or brought up some interesting questions? Now is your chance to find out more about your favorite science topics.

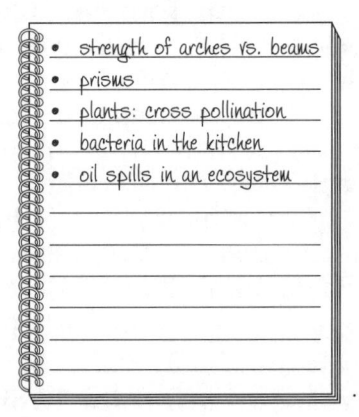

- strength of arches vs. beams
- prisms
- plants: cross pollination
- bacteria in the kitchen
- oil spills in an ecosystem

Solve a Mystery Another way to begin is to make a list of things that have mystified you. What makes a microwave oven cook some foods more quickly than a conventional oven? How do ants know to travel in a single line? Why are some bicycles faster than others? If the idea involves a mystery, chances are good that it would make an excellent topic for a science project.

The Library: Where the Topics Are Many kids get ideas from library books or even from the Internet. If you start researching early on in the process, you can get an idea of how much information has been written about your topic. Even though your goal is to design your own experiment, you're still going to have to research the topic. You will want to get some background information and determine how to go about testing your hypothesis. So don't be afraid of research. In the long run, it will actually make your project much easier for you.

> **TIP**
>
> An adult can help you think of an idea, but it is very important that your science project is clearly your own. At the science fair, your work will be compared only with that of other students close to your age.

From the list of topics that you brainstormed, choose five subjects that interest you most. Not all of these subjects have to be developed into an experiment; this is just a list of ideas of things that you may like to learn more about.

Investigative Question

When you've chosen five possible topics, you'll want to figure out what you want to know about each topic. Your investigative question is what you intend to find out during the course of your experiment. Using the topics you brainstormed earlier, you will choose a related question that your experiment will answer.

The Search Continues At the library, start by searching through the card catalog. (If you need help, ask your teacher or librarian. The faster you understand how to use the library's resources, the easier your research will be.) Look up key words from the subject that you are interested in. Looking up "oil spills" in the card catalog or *Reader's Guide to Periodical Literature* will lead you to a number of articles. If you have Internet access, you could try a search on "oil spills." You may include the word "penguin" in your search to find articles about how oil spills have affected penguins.

While you're at the library, try to think of two investigative questions that go with each topic you brainstormed. Record these ideas in your science project journal.

Here's an Example Let's say you decide to pursue the topic *oil spills*. You probably already know that large amounts of oil spilled into water can hurt the animals in an environment, but you want to know more. Maybe you want to know how an oil spill would affect a penguin's life span. Two investigative questions for this topic would be "How does oil in the water affect a marine ecosystem?" and "How do oil spills affect the life span of penguins?"

Choosing the *Best* Investigative Question The question "How does oil in the water affect a marine ecosystem?" is too broad to study in a short-term science experiment. Now consider the next question: "How do oil spills affect the life span of penguins?" The chances that you'll be able to do tests using penguins are very slim. You'll have to think of experiments that use resources available to you.

You don't have to throw out the topic if a particular experiment is not workable. Instead, think of a different testable question related to the same topic. For instance, you may want to learn more about how the ecosystem or food web of an area is affected when a large quantity of oil is introduced. This food web probably includes plant matter. Therefore, "How does oil affect plants?" is an example of a reasonable investigative question that you can test.

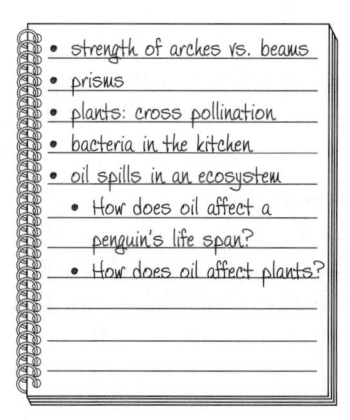

- strength of arches vs. beams
- prisms
- plants: cross pollination
- bacteria in the kitchen
- oil spills in an ecosystem
  - How does oil affect a penguin's life span?
  - How does oil affect plants?

Consult with Your Teacher

Now you have ten investigative questions, two for each of five topics. It will become very clear that some of the possible experiments you are interested in would be too involved to do in a science project that lasts only 6 or 8 weeks. Or, you may be able to think of a way to do an experiment but realize that the materials you would want to use would be too expensive. At this stage, you'll have a conference with your teacher about the investigative questions that you have. He or she will help you settle on one investigative question to use for your science project.

Avoiding Problems

Keep in mind the limitations of the science project. Ask yourself the following questions about your topic:

- Are the materials you need affordable?
- Are the materials you need available?

Consider the expense and availability of materials that you need for your project. For instance, unless you have access to an X-ray machine, you don't want to study how different materials affect the path of X rays.

- Does the science fair you are entering prohibit materials or specimens that you will need for your experiment?

Find out the limitations of the science fair itself. Ask your teacher about any restrictions on qualified science projects.

- How long will it take you to gather the necessary data?
- How much time do you have for data collection?

Be careful not to choose a topic that will take more time to investigate than you have to carry out the science project. For instance, you don't want to study the growth of a plant that grows only 2 cm a month.

Creativity Is the Key There are many ways to develop a workable project that is within your budget and still study a topic that interests you; it will just take a lot of creativity—and that's the key to a winning science project.

SCIENCE FAIR GUIDE **41**

Forming a Hypothesis

What Do You Think the Answer Is? Before you begin an experiment about oil spills, you would read about oil spills and guess that oil affects the health of plants in a negative way. It's typical scientific procedure to have an idea of the answer to your investigative question before you begin. In fact, the scientific method requires that you state a possible answer to the investigative question before you begin. This statement is called the hypothesis.

Some possible hypotheses for the investigative question "How does oil affect plants?" are

- "Oil does not affect the health of plants."
- "Oil negatively affects the health of plants."
- "Oil improves the health of plants."

What if the Hypothesis Is Inaccurate? It doesn't actually matter whether your hypothesis is accurate. The objective of a science project is to develop a hypothesis and then design a way to test it. In fact, few people will be concerned if the results of your experiment do not support your hypothesis. It is just as acceptable to have an inaccurate hypothesis as it is to have an accurate hypothesis—the challenge is in designing an effective experiment.

So take the plunge, and write down what you think will happen. After researching the example that we are using, you may come up with the following hypothesis:

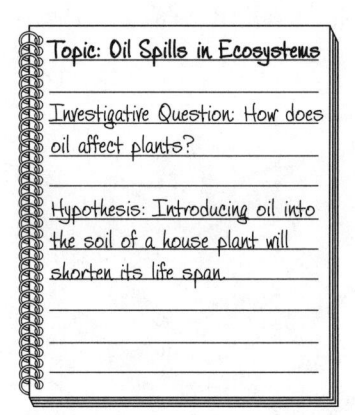

> Topic: Oil Spills in Ecosystems
>
> Investigative Question: How does oil affect plants?
>
> Hypothesis: Introducing oil into the soil of a house plant will shorten its life span.

✓ Make sure to check off the Phase 1 steps on your Student's Progress Report as you complete each one.

42 HOLT SCIENCE AND TECHNOLOGY

Phase 2—Research and Planning: Science Fair Success

> **DURING THIS PHASE, YOU WILL**
>
> **1.** research the hypothesis
>
> **2.** change or reconfirm the hypothesis with the teacher
>
> **3.** contact all appropriate people
>
> **4.** record all research sources in the science project journal
>
> **5.** fill out the Procedural Plan for Action and obtain signatures
>
> **6.** develop the initial plan for display materials

Remember to update your Student's Progress Report as you go along. You will have until _____ to finish Phase 2.

Researching and Planning

Doing further research will help you decide what your experiment will be. Remember to write all of your research notes in your science project journal.

Imagine for a moment that your hypothesis is "Bridges with arches provide more support than simple beam bridges." For your project, you want to build two models: one testing the forces that bridges with beams can withstand and another testing the forces that bridges with beams *and* arches can withstand. Before you can really do this, you will have to read about the design of bridges.

Ensuring Success By thinking ahead, you can prevent problems from popping up later. So start planning now, and you'll be much more likely to end up with a successful project. Now's the time to

- make sure you have enough time to complete your experiment.
- make sure you can cover the costs involved.
- start thinking about the science fair display.
- ask others for any information or help you may need.

Is This the Project for You?

After additional library research, you'll check in with your teacher to reconfirm your hypothesis. By now, you probably have a good feel for how much information has been written on your topic. You may also have some good ideas about what you can do to test your hypothesis. If you need to change your topic at this point, you can still do so. If you feel that you won't have enough time or money to complete your experiment, speak to your teacher. Remember, this is your last chance to change topics! After this point you will be too far along in the timeline to safely change your plan of action.

Further Research

You have already begun researching your topic by the usual means, such as reading books and checking out the Internet. Perhaps you haven't thought about the following ways of gathering information:

Interview an Expert You might want to interview an expert in the field you're studying. If you are interested in studying sedimentary rocks, for example, you could talk to a geologist. A geologist could probably give you some good tips on where to find the materials you need. Remember to record the details of the interview in the bibliography section of your journal. You'll need to give the expert credit for the information that he or she has given you.

Let the Letter Ask Writing a letter is one of the best ways to get information. It sometimes takes a while to receive a response, though, so write as early as possible. Let's say you were interested in finding out how to dispose of paint and other chemical wastes without hurting the environment. You might write to someone at a paint company. You can ask for information in a letter like this one:

Green Paint Company
4500 East South Street
Detroit, Michigan 70555

Dear Sir or Madam:

I am a seventh grader preparing for the Springfield School District Annual Seventh-Grade Science Fair. The purpose of my experiment is to see how proper disposal of leftover paint can prevent heavy metals from contaminating drinking water supplies.

Could your company provide me with a list of chemical ingredients contained in your product? I would also appreciate some guidelines on how to safely dispose of leftover paint.

I hope you will be able to help. Thank you for your time and effort.

Sincerely,

Meredith Phillips

---

44 HOLT SCIENCE AND TECHNOLOGY

Thank Others for Their Help When you receive a response from a company, transfer any useful information into your science project journal. It's also a good idea to keep the actual letter in a safe place so that you can refer to it again later. Remember to thank anyone who helps you by supplying information or materials.

Citing All Your Sources

Your sources may include books, magazines, newspapers, Web sites, television programs, videos, or even interviews with live people. You will need to include all of these sources in the bibliography for your research paper. Your science project journal is the perfect place to keep track of this information.

How to Cite a Reference In your science project journal, record the title, author, publisher, and copyright date of each source that you use. If you perform an interview, record who you spoke to, what you discussed, and when and where the interview took place. You can keep all this information organized by devoting a few journal pages to your notes on information sources.

Preparing to Conduct the Experiment

You have a hypothesis that you are happy with, and you are learning more about your topic. Now it's time to consider a procedural plan for your experiment. Most scientific experiments follow the same basic rules, which are explained below.

Being in Control When testing your hypothesis, you'll want to establish a *control group* and one or more *experimental groups*. The control group and the experimental groups are exactly the same except for one factor, which varies in the experimental groups. The factor that differs is called the *variable*. Because the variable is the only factor that differs between the control group and the experimental group, scientists know that this factor is responsible for the results of the experiment.

If you were planning to test the effects of an arch on the strength of a bridge made of beams, you must test the strength of a bridge that doesn't have an arch and the strength of the same bridge with an arch. Bridges that have no arch (beam bridges) are in the control group. Bridges that have beams *and* arches are the experimental group. The variable is whether the bridge has an arch.

Then if you wanted to test the effects of suspension cables on the strength of a bridge, you'd use the same control group as before—beam bridges. Your experimental group would include beam bridges with suspension cables. The new variable is whether the bridge has suspension cables.

It Depends The physical structure of a bridge—whether it has an arch—can differ from one bridge to the next. Since the bridge's physical structure does not depend on any other characteristic, it is called the *independent variable*. The independent variable can be changed by the researcher.

SCIENCE FAIR GUIDE **45**

The strength of the bridge *depends* on whether the bridge has an arch, so the strength of the bridge is the *dependent variable*. When you measure the dependent variable you are measuring the result that occurs when the researcher changes the independent variable. The dependent variable is what the experiment is actually measuring.

There Can Be Only One It is very important that you have only one variable. Otherwise you are not effectively testing your hypothesis. For instance, your test of the bridges' strengths would be influenced by the materials from which the bridges were made, so all bridges in your experiment must be made of the same materials. *All factors of the control group and the experimental group(s) must be exactly the same except for one variable.*

> **Experiments Versus Demonstrations** When you don't know what you want to do for an experiment, it can be tempting to demonstrate a concept that scientists have already figured out. As a researcher, you will focus on performing an experiment instead of a demonstration. Though modeling a volcano's eruption is an interesting demonstration, it is *not* an experiment. Figuring out how altitude affects the boiling point of water *is* an experiment. Ask your teacher if you need help developing an experiment that tests your hypothesis.

The Procedural Plan

Look on the next page to find a blank form for your Procedural Plan for Action. You can use this worksheet to write down your ideas for your experiment. It can help you think through the details of your science fair project.

When completing the procedural plan, it is important to begin planning for the display. Your teacher will provide you with a list of rules for the science fair you are going to enter. The rules will include important information about display sizes and materials. You will also need to begin gathering the materials that you will need for the experiment and the science fair.

Put Your Ducks in a Row Some students find it helpful to use a schedule or a list to organize all the work that they need to do. You can use the Task List (page 48) to create your own personal schedule. Fill out your task list however you like. You can use it as a personal to-do list or as a place to write down which days you want to go to the library or meet with your teacher.

✓ Make sure to check off the Phase 2 steps on your Student's Progress Report as you complete each one.

Name _____ Date _____ Class _____

Procedural Plan for Action

Describe your experiment's procedure.

What will be your control group?

What will be your experimental group?

| What is your dependent variable? | What is your independent variable? |

What kind of location or setting will you need for your experiment?

What kind of materials will you need for your experiment?

What costs do you expect?

Parent Signature _____ Date _____

Teacher Signature _____ Date _____

SCIENCE FAIR GUIDE

Name _____ Date _____ Class _____

Task List

Task to be accomplished	Goal date	Date accomplished

Phase 3—Data Collection and Analysis: Science Fair Success

> **DURING THIS PHASE, YOU WILL**
>
> 1. conduct the experiment safely
> 2. choose an appropriate sample size
> 3. perform several trials of the experiment
> 4. collect data accurately
> 5. record all data and observations in your science project journal
> 6. graph or chart the data and look for trends
> 7. prepare a written conclusion supported by the data

Before You Begin

- **Have your teacher or parent approve your experiment.**
- **Read over your Safety Contract.**
- **During the data collection, make sure that there is an adult present.**

Your teacher will schedule some in-class data collection days, so remember to take the necessary materials to class. Don't forget to update your Student's Progress Report as you go along. You will have until _____ to complete Phase 3.

Types of Data

Data can take two different forms: data can be *quantitative* (a value that can be measured or counted) or *qualitative* (a value that can be described but cannot be measured or counted). Some projects may combine both forms of data.

Quantitative Data: Count It Up *Quantity* and *quantitative* have the same word root. Just remember that quantitative data have to do with numbers or quantities that you can measure. Examples of quantitative data are the number of bird chirps that you hear on a cold day or the width of a layer of rock in a cliff wall.

Qualitative Data: Spell It Out The word *quality* is related to the word *qualitative*. Taking qualitative data means that you will be describing your observations with adjectives instead of numbers. Examples of qualitative data are descriptions of the color and shape of the rock in each layer of a cliff wall. Drawings and photographs are also qualitative data.

Whichever kind of data you have decided to collect, *remember to write it down.* Be certain to record *all* results of your tests in your science project journal. Recording everything as it happens not only will help you keep your information in one place but also will make it easier for you to avoid errors.

SCIENCE FAIR GUIDE **49**

Sample Size and Multiple Trials

Test More Than One Subject *Sample size* is the number of subjects you test. Your sample size must be large enough to allow you to draw accurate conclusions from your data.

For example, if you were comparing the differences in hand-eye coordination between 20-year-old and 40-year-old women, the data taken from just one woman in her 20s may not give you accurate results. She could be the fastest woman in the world! Testing only her, you would conclude—possibly incorrectly—that all women in their 20s are faster than all women in their 40s. If you tested several other women in their 20s and compared this data to the data taken from several 40-year-old women, you would have a more realistic picture of the actual trends among 40-year-olds and 20-year-olds.

Record in your journal the number of subjects that you tested. Also record any details that might affect your results. For instance, in the example above, you would record each subject's age, sex, height, weight, the time of day the subject was tested, the number of times you tested each subject, etc.

Play It Again, Sam! When you are conducting an experiment, it is necessary to do *multiple trials*. This means you should perform each test several times. For instance, if you are going to compare your body temperature in the morning with your body temperature in the evening as part of your project, make sure that you test your temperature on several days before you try to draw any conclusions. And if one day's temperature differs wildly from the rest, you might consider that there was an error and try to figure out what went wrong.

Taking Accurate Measurements

Scientists need to be as exact as possible in taking measurements. It's almost impossible to measure something exactly, so scientists usually measure something more than once and then use the average of the results. This approach helps to account for the uncertainty of each individual measurement.

Always double-check the measurements you take before you record them in your journal. Measure carefully and make each measurement level. For instance, if your experiment requires the use of measuring spoons or cups, be consistent: don't fill some of them so that they are heaping and others so that they are not quite full.

Go Global Most scientists use the International System of Units (SI) in all their work. The International System of Units is a global measurement system that helps scientists share and compare their observations and results. It is usually best to use this system, and some science fairs require that all measurements be expressed in SI units. Your teacher has a unit conversion chart for common measurements of length, mass, volume, and temperature that may be useful.

Creating Charts and Graphs

As you are collecting your data, you may want to keep in mind that you will be required to display your results and conclusions at the science fair. You can make it easy for people to understand the relationship between your variables by displaying your data in a chart or graph. It may be useful for you to make the first drafts of your graphs in your journal. This will let you decide the best way to illustrate your data before you make the final graphs for your display.

> **TIP**
>
> When you create a graph, make sure that you leave equal spaces between the numbers on the axes and that you number the axes consistently. For instance, if you start with the number 0 and the next values are 5 and 10, you can't skip to 20. The next number would have to be 15.

Bar Graphs

Use a *bar graph* if you want to compare different types of data. In the case of a bar graph, each bar represents a group of data.

- It is important to make it easy to identify each bar. For example, you could choose to use polka-dotted bars for one group of data and striped bars for another.

- Every graph always needs a key so that people can easily tell what each color or pattern represents.

- Make each key the same so that graphs are easy to compare. That way, if you had measured data on three different days, a person could quickly distinguish between the groups of data.

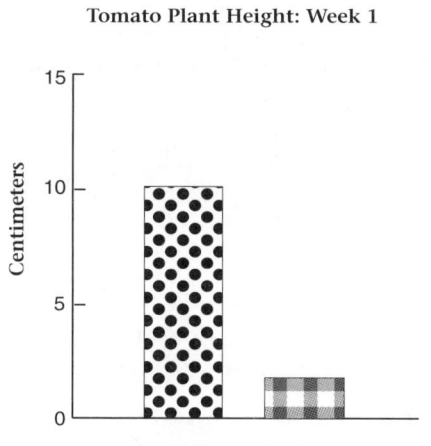

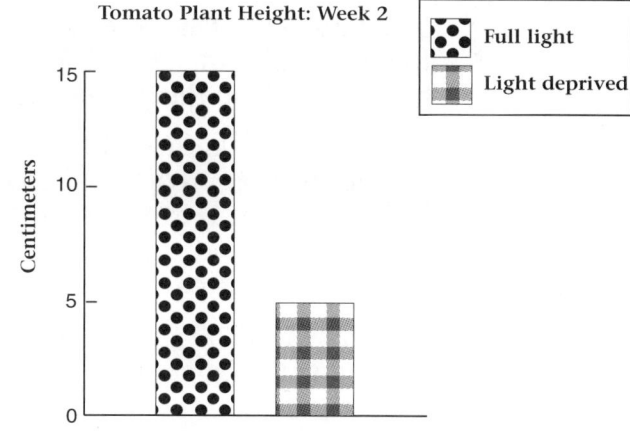

SCIENCE FAIR GUIDE **51**

Line Graphs

Use a *line graph* if you want to show how the dependent variable is affected by changes in the independent variable or if you want to show how data change over time.

- In a line graph, place the dependent variable on the vertical (up-and-down) axis, or the *y*-axis. The independent variable should be on the horizontal (left-to-right) axis, or the *x*-axis.

- Plot your data as carefully as possible. Then connect the points.

- If you decide to record the results of more than one experiment on one line graph, you may choose to use a different color of ink for each set of points (and the connecting line). Be sure to include a key explaining the colors.

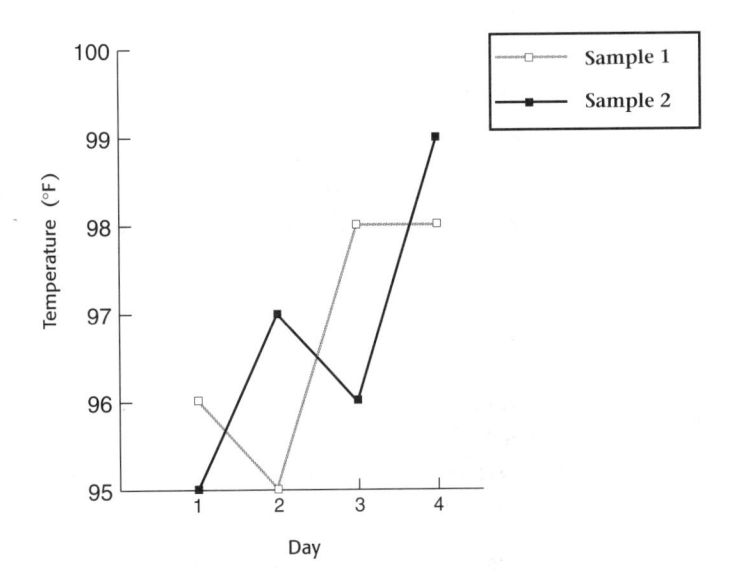

Tips

This is where graph paper will come in handy. Your graph will be much neater and easier to draw if you use graph paper. You will find a sheet of graph paper on page 54.

Pie Charts

Showing percentages is easy to do with a *pie chart,* a round chart that looks like a sliced pie. You can quickly see which group has the biggest slice and therefore contains the most data.

- The size of a group's slice indicates the proportion of the whole that the group represents. Say you want to show that 30 percent of people in an experiment sneezed when exposed to a bright light and 70 percent of people didn't sneeze. You would use 30 percent of the circle to represent the sneezers.

- A circle has 360°, so you would multiply 0.30 by 360° to get the number of degrees that should be used to represent the sneezers.

$$0.30 \times 360° = 108°$$

Sneezing After Exposure to 10 Minutes of Bright Light

30% of people sneezed

70% of people didn't sneeze

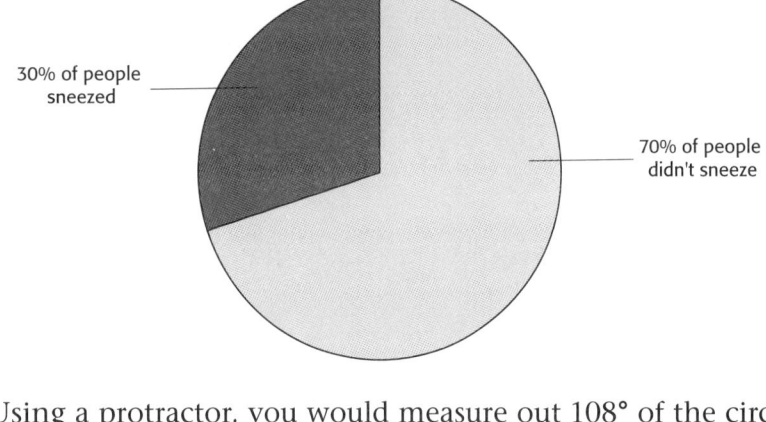

- Using a protractor, you would measure out 108° of the circle for the sneezers. The other part of the pie chart (252°) would be reserved for the nonsneezers.
- Make each section of the chart a different color and include a key or labels to make the graph easy to understand.

Analyzing Data and Drawing a Conclusion

After you've gathered all of your data, you'll want to analyze your results. In the analysis, ask yourself, "What are the data telling me? What trends do I see in my graphs? Are the data for the control group different than the data for the experimental group?" Write your analysis in your science project journal.

What Does It All *Mean*? If your results are mathematical, it will help you to understand the concepts of *mean* and *median*. Mean is the average of your data, and median is the middle-most value when all measurements are listed in order from smallest to largest. Two experiments may have the same average result but differ in how the results are distributed. Compare the means and medians and see how they differ. Ask your teacher or your parents for help with mathematical concepts that you aren't sure about.

Drawing Conclusions

The main question you should ask yourself when drawing a conclusion is, "Do my results agree with my hypothesis?" If they do, why do you think they do? If they don't, how are they different? And, why do you think they differ? Remember that it is *not* important for the hypothesis to be correct. It is important, however, that you explain *why* you got the results you did. Write your conclusions in your science project journal.

Be sure to mention in your conclusion what factors you believe contributed to your results. Then, briefly explain possibilities for new experiments that would control these factors. Also, mention any investigative questions that came up during the experiment. These questions will guide other researchers who find your results interesting and want to study the topic more.

✓ Make sure to check off the Phase 3 steps on your Student's Progress Report as you complete each one.

SCIENCE FAIR GUIDE

Name _____ Date _____ Class _____

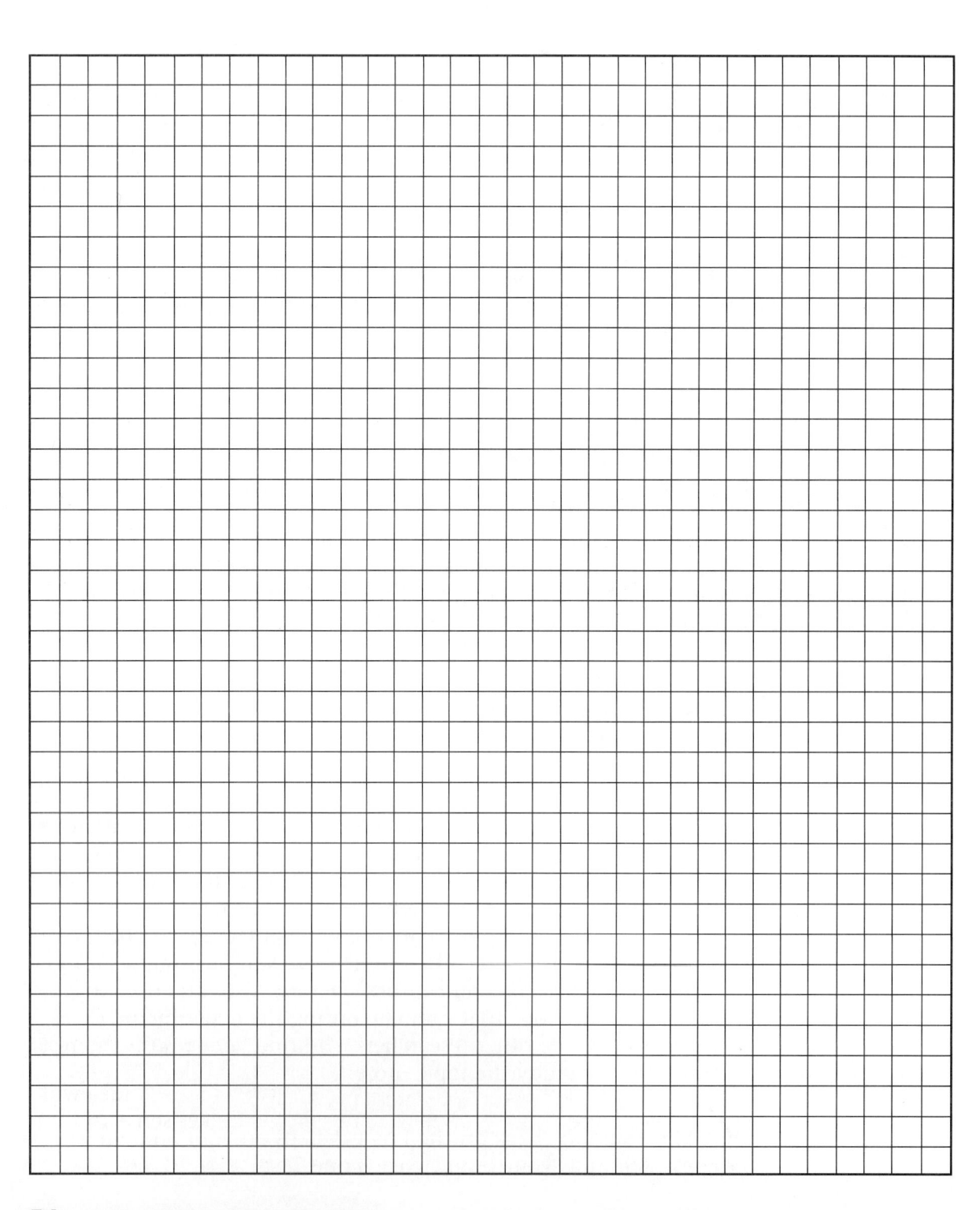

Phase 4—Writing a Report: Science Fair Success

DURING THIS PHASE, YOU WILL

1. answer the questions on this page
2. prepare an outline and discuss it with your teacher
3. prepare a draft and discuss it with your teacher
4. revise the draft according to your teacher's feedback
5. turn in a completed draft

All of the time you've spent recording every little detail in your science project journal is about to pay off. Your written report will represent your ideas and conclusions about your project, so you'll want to make sure that it is well thought out and neatly prepared. Don't forget to update your Student's Progress Report as you go along. You will have until _____ to complete Phase 4.

Putting Your Ideas on Paper

Before you begin, answer the following questions in your science project journal. When you are through, you may have a better idea of how to start on your written report.

1. *How did you first decide on your idea?*
2. *What was your favorite aspect of the experiment?*
3. *What was something new that you learned?*
4. *What was something unexpected that happened?*
5. *What were the ups and downs of the whole process?*
6. *What did your data show?*
7. *What would you do differently next time?*

TIP

Have fun with your report. You've already done most of the work. Now, just carefully describe what you did in each phase, and explain every detail of your experience.

Creating an Outline

Now that you've answered the questions on this page, you are ready to make an outline of your report. An outline is a framework of what is going to go inside a report. Most scientific reports follow the same order as the steps of the scientific method, explaining the entire process from beginning to end. Using index cards for each idea will allow you to make sure that all the information makes sense in the order that you are going to tell it.

SCIENCE FAIR GUIDE **55**

Making a List Include at the beginning of the report any background information that a reader would need to understand your project. Then you will state the purpose and hypothesis of your project. Briefly describe your procedure and the data that you acquired. Finally, you'll want to illustrate your conclusion using charts, graphs, or photographs of your data. You may want to include your answers to the questions on page 55.

> **TIP**
>
> You may choose to use some of your charts or graphs in your report. The best figures to use are those that clearly show the trends you found in your data. Always title and label your figures, and, if possible, write a sentence telling what they illustrate.

Checking It Twice When you are done, have your teacher check your outline. Your teacher will make some suggestions about how to improve your report. Pretend the report is not your own, and try to see how your teacher's suggestions could make it even better. Professional scientists and writers get constructive criticism about everything they do before they show their results to the world. It is an important part of the writing process, and it will help you improve your report so that it is ready to impress the science fair judges.

Writing the Draft

After you've thought about your teacher's suggestions, you may want to change your outline. When you are happy with your outline, you can start a draft of the written report.

> **TIPS**
>
> The following are some helpful writing tips:
>
> - Before you begin, make sure that you know how many pages your report should be.
>
> - As you write, you may ask your parents to read sections to make sure that you are on the right track.
>
> - Show your finished draft to your teacher. Your teacher will make comments to help you create the best project that you possibly can.
>
> - Don't forget about neatness and spelling! Judges will notice if you have not been careful in your work.
>
> - Finally, remember to complete a bibliography of the sources you used. Be sure to give credit to the people who helped you in your project—your teacher, your parents, professional scientists, or others.

✓ Make sure to check off the Phase 4 steps on your Student's Progress Report as you complete each one.

Phase 5—Creating and Exhibiting a Display: Science Fair Success

> **DURING THIS PHASE, YOU WILL**
>
> 1. sketch possible designs for the display
> 2. create a display board within the appropriate parameters
> 3. display results in a clear and interesting manner
> 4. give an oral presentation as practice for the science fair interview

After this phase, you can tell everyone that you know what it is like to be a real scientific researcher—and mean it! Don't forget to update your Student's Progress Report as you go along. You will have until _____ to complete Phase 5.

Before You Begin

Before you begin your display, review the science fair display and interview guidelines. Some science fairs have a list of items and materials that you are not allowed to use in a display. Double-check to be sure you have followed these guidelines. You wouldn't want to be disqualified after all of the hard work you've already done.

The Layout of Your Display

Displays are usually divided into three sections. You can bend a cardboard box so that it has a large middle section and two smaller "wings" that fold inward across the center. This design allows your display to fold flat, making it easier to transport and store. You can purchase hinges at an art supply or hardware store if you prefer to attach the "wings" that way.

SCIENCE FAIR GUIDE **57**

An Example The illustration below is one example of how information can be laid out on the display. You can do it differently, but remember to place the information from left to right in the general order that you performed each item. It is also common to place models, samples, demonstration props, or small pieces of equipment in front of the display board.

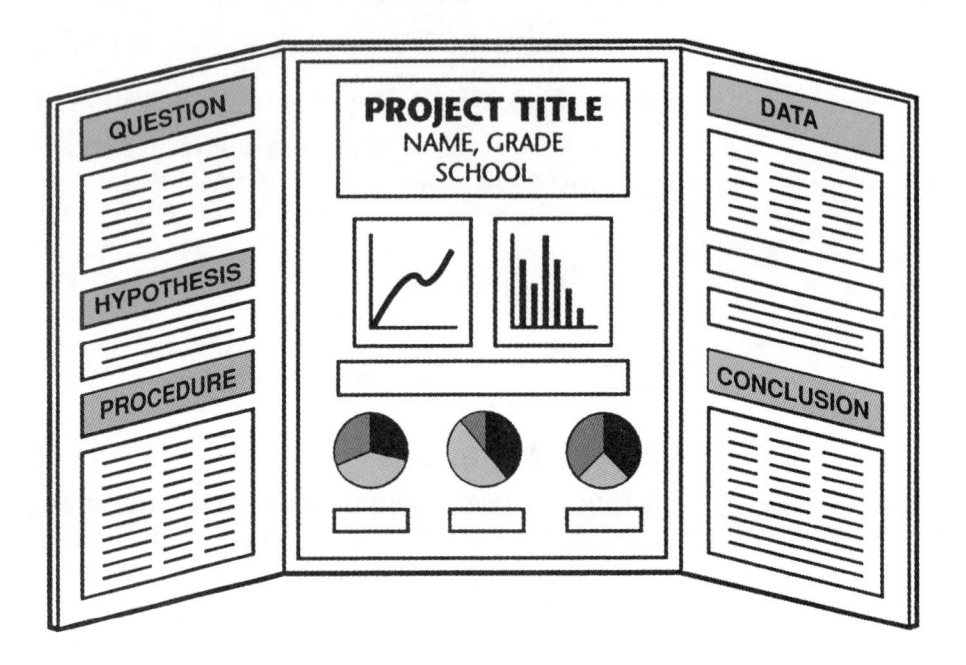

Here's how the information is organized on the display shown above:

- **Top Left** This section provides basic background information and introduces the purpose and hypothesis of the project.

- **Bottom Left** This section briefly explains the procedure that was followed (review your Procedural Plan for Action).

- **Right-hand Panel** Brief written summaries of the data and the conclusions are located on this panel. The research is displayed so that it is obvious that the data support the conclusions.

- **Center** The middle panel contains the title of the project and the name, grade, and school of the researcher. Charts, graphs, photographs, and other illustrations are displayed here.

- **Keep It Simple** The display touches on all aspects of the project, but keeps the information general. The details of the project belong in the written report.

Designing Your Display

Back to the Drawing Board Before you construct a display, sketch some ideas of how you want your display to look. Sketching it out on paper lets you easily choose colors, borders, sizes, lettering, and even arrangement of items in your display.

Materials Most students will use corrugated cardboard, cork board, or foam core to construct their display. You can recycle by calling a local appliance store to get a large, corrugated box from a refrigerator, washing machine, or TV. Observe appropriate safety precautions and make sure that an adult helps you cut the cardboard to regulation size.

Remember—Neatness Counts! There may be requirements about the lettering for the display. If you write the information for your display by hand, make sure the writing is neat and easy to read. Your main title and major subtitles should be readable from a distance, and any other information can be smaller. If you use paper or plastic lettering or stencils, use a ruler to apply them in a straight line.

Creative, Yet Clear While you want your display to be as interesting as possible, the design should not distract from the content. Be creative with borders, font, and layout, but make sure that a judge would find it easy to read the information contained in the display. Illustrations should be informative, not just decorative. In your display, you want to impress the judges with the project's seriousness yet be unique and have some fun.

> **TIP**
>
> A simple display is best if it clearly shows what you have learned. An expensive computer display or a flashy presentation is useful only if it relates to your results and helps make your conclusions more clear.

The Interview

If you are going to make a presentation to the judges, it's in your best interest to practice at least once. Practice explaining your project to someone who knows nothing about it. Your family and your classmates are good audiences. It may be difficult at first, but once you run through it a few times, you'll have a great advantage over students who haven't bothered to practice. If you're part of a group project, each person in the group should be responsible for presenting a certain aspect of the project, such as the purpose, hypothesis, or conclusion.

Summarize It! You'll want to prepare a summary of what you did. You can do this by following the steps below.

- Explain why your subject interested you.
- Define the hypothesis you developed.
- Describe how you decided what type of data to collect.
- Outline your Procedural Plan for Action.
- Summarize the data you actually collected.
- Explain the conclusions you drew after you analyzed the data.
- Describe to the judges what you would do differently if you had another chance, and tell them why.

Tricks of the Trade Here are more suggestions that may help you during the interview:

- Carry an index card with an outline of what you want to say, and refer to it if you forget something during your interview.

- Don't read to the judges from your report or from notes—they would rather hear you speak naturally.

- Have copies of the written report near your display so that interested people can learn more about your project.

- Offer a copy of your report to the judges so that they can read about what you have done.

- If a judge asks you a question that you are unable to answer, stay calm. Explain that you aren't sure about the answer to that question, and offer to explain a part of the project that you're more comfortable with.

- If the judge offers you suggestions or says that he or she might have done something differently, try not to react angrily. The judge is only trying to help you be a better scientist.

- It is perfectly natural to be nervous—even seasoned scientists get the jitters. Just remember that this is your project and you are the best person to explain it to others!

✓ Make sure to check off the Phase 5 steps on your Student's Progress Report as you complete each one.

Part 3: Resources for Parents

Welcome to the World of Science Fairs!

The *Science Fair Guide* is designed to direct teachers, students, and parents through all phases of a science project, preparing a student to participate in a science fair. A science project serves more than one purpose. The most obvious goal is to extend a student's understanding of science. Science is hands-on by nature, and there is no doubt that hands-on experiences facilitate the learning process.

The purpose of Resources for Parents is to help you guide your student through a science project using the following five phases, which mirror the scientific method:

- Phase 1: Generating an Idea
- Phase 2: Research and Planning
- Phase 3: Data Collection and Analysis
- Phase 4: Writing a Report
- Phase 5: Creating and Exhibiting a Display

The Scientific Method The scientific method allows scientists to further the advancement of knowledge in a strategic and unified manner. When students use the scientific method to design and execute a project they use the same steps that professional researchers use to glean new information about the world.

THE SCIENTIFIC METHOD

The steps of the scientific method that your child will be using appear below. Each step is explained in greater detail in Phases 1–5 of Resources for Students.*

Purpose: developing an investigative question

Hypothesis: making an educated guess about the answer to the investigative question based on research

Experiment: testing the hypothesis, collecting data, and making observations

Analysis: organizing data from the research and experimentation and looking for patterns

Conclusion: determining if the hypothesis is supported or disproved by the experimental results

Communicating the Results: sharing the conclusion with others

*In Chapter 1 of your child's textbook, these steps are called Ask a Question, Form a Hypothesis, Test the Hypothesis, Analyze the Results, Draw Conclusions, and Communicate Results.

Building Scientific Confidence Most importantly, science projects make science more fun and relevant to the student. Every student, regardless of aptitude, can benefit from planning and executing a science project. A science project can be a great way to spark a student's interest in science or to help a student develop a broader interest in research.

Philosophy of Science Fairs

Science fairs range in scale from a single class of students showing one another their projects to an international event offering scholarship money as prizes. From the largest science fair to the smallest, science fairs generally share a common philosophy and similar values:

- emphasis on using the scientific method
- investigation by experimentation (inquiry-based learning)
- development of critical thinking skills
- opportunity for a positive learning experience
- extension of formal science education

How to Help as a Parent

It is up to the student to decide what to study. You can help by motivating your child and listening to his or her ideas. However, it is crucial to remember that it is up to your child to design and execute the entire project. Judges at a science fair take particular care to note that the work is the student's and that the student understands the topic, the research, the experiments, the analysis of data, and the conclusion. The judges expect that the student has received some help from another person, such as a parent or teacher, and that such help will be credited in the display.

Support the Troops Your child may need more attention than a teacher can give to each student in a large class. Some class time will be devoted to researching at the library; however, your child may find it helpful to do more library research outside of school. Expect your child to spend time brainstorming, researching, planning, experimenting, analyzing data, writing a report, and constructing a display. You may offer to spend time with your child at the library. You can also help by encouraging your child to record everything in his or her science project journal, such as notes from brainstorming, sources used during research, and observations made during data collection.

Your child has been told that an adult must be present during all data collection. Please supervise the experimental phase for safety purposes. You may refer to the Safety Guide to help avoid accidents during data collection.

SCIENCE FAIR GUIDE **63**

When You Should Help It is very easy to take control of a student's project, especially if you think it should be done differently. Remember that this project is a learning experience for your child, and he or she will not benefit from a project performed by you. If your child is performing all the necessary tasks to an acceptable standard and is not requesting assistance, your job is to supervise. If your child asks for help, appears to be struggling, or is performing below acceptable standards or with disregard for safety measures, then you may wish to offer assistance.

Judging

The criteria on which a science fair is judged can vary, and most judges evaluate projects using the following criteria:

- use of scientific thought
- creativity of approach
- thoroughness of investigation and research
- skill of experimental technique
- clarity of expression in presentation and report

Besides assessing the five categories that appear above, some science fairs reward memorable presentations or displays.

Student Understanding It is very important that the student's work be evident in the project. The scientific process and the manner in which the research was conducted tend to outweigh the actual display of information. The judges will be looking carefully to see that the student has an understanding of his or her project and is responsible for the final product. In a group project, the judges will be looking for evidence that each group member completed specified tasks that furthered the progress of the group.

Enclosed Materials

The handouts for each of the five phases will help you guide your son or daughter through each phase of the project. In preparing for the science fair, please understand that this project is a fun opportunity for independent learning rather than for competition to identify winners and losers. Coming away with a new interest in a particular subject or a new understanding of a scientific principle can be more rewarding than a prize.

A Parent's Progress Report is included in Resources for Parents on pp. 65–66. Please refer to this sheet throughout the project to make sure that your child is on track according to the timeline enclosed in the parent letter. The teacher may use the Parent's Progress Report for grading purposes.

Name _____ Date _____ Class _____

Parent's Progress Report

❏ individual project   ❏ team member

For each step of each phase of the science project, record the task due date and the date the student accomplished the task. You might want to reward your child for completing each phase.

Phase 1—Generating an Idea	Date due	Date accomplished
Student brainstormed five possible subjects.		
Student came up with two investigative questions for each topic.		
Student consulted with teacher and parents about project possibilities.		
Student chose a suitable topic.		
Student formed a hypothesis.		
Student discussed topic and hypothesis with teacher and gained approval.		
Student recorded ideas in the science project journal.		

Phase 2—Research and Planning	Date due	Date accomplished
Student researched the hypothesis.		
Student reconfirmed or changed the hypothesis based on further research and then gained teacher approval.		
Student contacted all appropriate people before beginning data collection.		
Student recorded all details of research so far in a bibliography in the science project journal.		
Student filled out the Procedural Plan for Action and obtained necessary signatures.		
Student developed the initial plan for display materials.		

Parent comments: _____

_____

_____

SCIENCE FAIR GUIDE **65**

Name _____ Date _____ Class _____

Parent's Progress Report, continued

Phase 3—Data Collection and Analysis	Date due	Date accomplished
Student conducted the experiment safely.		
Student chose an appropriate sample size.		
Student performed several trials of his or her experiment.		
Student collected data accurately.		
Student recorded all data and observations in the science project journal.		
Student graphed or charted data and looked for trends.		
Student prepared a written conclusion supported by the data.		

Phase 4—Writing a Report	Date due	Date accomplished
Student answered the questions on page 55.		
Student prepared an outline and discussed it with the teacher.		
Student prepared a draft and discussed it with the teacher.		
Student revised the draft according to the teacher's feedback.		
Student turned in final draft of the written report.		

Phase 5—Creating and Exhibiting a Display	Date due	Date accomplished
Student sketched possible designs for display.		
Student created a display board within the appropriate parameters.		
Student displayed the results in a clear and interesting manner.		
Student gave an oral presentation as practice for the science fair interview.		

Phase 1—Generating an Idea: Getting Involved

> **DURING THIS PHASE, STUDENTS WILL:**
>
> 1. brainstorm five possible subjects
> 2. come up with two investigative questions per topic
> 3. consult with teacher and parents about project possibilities
> 4. choose a suitable topic and investigative question
> 5. form a hypothesis
> 6. discuss topic and hypothesis with the teacher and gain approval
> 7. record ideas in their science project journal

At the beginning of the project, you may wish to consult the timeline that was included in the packet of information that was sent to you. Also, you may want to contact the teacher to find out the specific judging criteria and display requirements for the particular science fair that your child will be entering.

This is fun! Throughout the course of the project, especially during the beginning stages, you may remind your child that one of the main goals of the project is to have fun. While healthy competition can be exciting, the purpose of a science project is to learn science through inquiry in a hands-on experiment.

The Journal: Your Child's Most Important Tool Your child will record everything he or she does in his or her science project journal. This includes all brainstorming lists, research notes, data, and observations. Your child will need the journal for all phases of the science project, so it is important that he or she keep careful track of it. Also, you may want to check the journal periodically as a way of keeping up with your child's progress.

Brainstorming for Topics

During Phase 1, students will be brainstorming for scientific subjects that interest them. Subjects can be as simple as how light affects plant growth or as complex as how carotenes affect cancer growth. Students often find the pressure of coming up with a suitable idea the most challenging part of a project. You can help by acting as a sounding board for ideas or reminding your child of things that he or she has been curious about around the house. Students will need to brainstorm five possible topics and eventually develop two investigative questions for each topic.

The Investigative Question

The investigative question is what the experiment is intended to find out. This question should narrow down the topics to specific areas of interest. For example, if a student chooses biological clocks as a topic, he or she might ask, "Does the length of time for which a hamster is exposed to daylight affect how much it eats?"

Ten Questions After every student has selected five topics, the class will visit the library to research the subjects that students have chosen. Students will create 2 investigative questions for each topic, a total of 10. It is likely that one of these questions will develop into a successful project. However, brainstorming more than one topic will allow students to change topics with little difficulty if they choose a topic that doesn't prove fruitful.

Choosing the *Best* Investigative Question As it turns out, "Does the length of time for which a hamster is exposed to daylight affect how much he eats?" may be a problematic investigative question because many science fairs prohibit the use of vertebrates in any experiment. The teacher will be familiar with the specific rules of the science fair and will steer students away from experiments that would lead to disqualification. There are many testable investigative questions within a topic, and one of them will probably fit the requirements of the science fair. Another investigative question for the subject "biological clocks" that may be more appropriate for a science project is "Does the length of time for which a flowering plant is exposed to daylight affect the rate at which it produces flowers?"

The Hypothesis

Once a student has selected an investigative question, the next step is for him or her to develop a hypothesis. A hypothesis is an educated guess that answers the investigative question. After reading about flowering plants and their growth patterns, the student will be able to guess whether flowering plants have biological clocks. Two appropriate hypotheses for that investigative question would be as follows:

- The length of time for which a flowering plant is exposed to daylight affects the rate at which it produces flowers.

- The length of time for which a flowering plant is exposed to daylight does not affect the rate at which it produces flowers.

Have your child list all the possible answers to the investigative question and choose the one that is most likely to be correct according to the research he or she has done. The chosen answer will serve as your child's hypothesis.

What If the Hypothesis Is Inaccurate? Often students feel great pressure to be right about things. If you find that your child is discouraged because the results of his or her experiment disprove the hypothesis, remind him or her that the quality of the project is independent of the accuracy of the hypothesis.

✓ If your child is involved in a group project, try to monitor his or her individual participation in the project. Consult the timeline, and check your child's progress at the end of Phase 1.

Phase 2—Research and Planning: Getting Involved

> **DURING THIS PHASE, STUDENTS WILL**
>
> 1. research the hypothesis
> 2. change or reconfirm the hypothesis with the teacher
> 3. contact all appropriate people
> 4. record all research sources in the science project journal
> 5. fill out the Procedural Plan for Action and obtain signatures
> 6. develop the initial plan for display materials

Research

By now your child has settled on an investigative question and a hypothesis, and it is time to do further research. Your child may need help with basic researching techniques both at the library and on the Internet. Further library research trips may be necessary for your child to confirm his or her hypothesis and determine how to test the hypothesis effectively. If your child has access to the Internet at home, you may want to help him or her find appropriate sites for science project research. Please continue to encourage your child by discussing the project with him or her.

Your child may wish to obtain information using other methods, such as seeking an interview with a scientist or writing a letter to a company. Encourage your child to contact all sources early in Phase 2 in order to allow plenty of time for a response.

Bibliography Students will need to prepare bibliographies for both the research project and the display, so remind your child to keep careful note of all sources. These include documentaries, interviews, Web sites, television programs, books, newspapers, and magazines. Your child is expected to record this information in his or her journal.

Procedural Plans

After a few trips to the library and some interviews or Internet research, your child will know whether he or she is comfortable with the chosen investigative question. Then, your child will consider how he or she is going to test the investigative question. Students' packets include discussions of control groups versus experimental groups and dependent variables versus independent variables. The teacher will most likely discuss these concepts in class.

Voice Your Concerns The teacher has required that your child fill out the Procedural Plan for Action on page 47 of Resources for Students. Students are also encouraged to use the Task List (page 48). The teacher will be watching for projects that require too much time or money. *If, for some reason, you think that the project is unreasonable (because of cost, workload, or lack of interest on your child's part), now is the time to contact the teacher. Soon it will be very difficult for your child to switch topics.*

SCIENCE FAIR GUIDE **69**

The Price is Right In most cases, there will be minor costs involved in your child's science project. As your child formulates a science project, he or she will be asked to evaluate the costs involved with performing each experiment. You can assist your child with this process and help create an informative project that stays within your family's budget. In certain cases, a science fair entry fee or travel expenses may be required. In general, here are a few items you can expect your child to need:

- journal
- materials necessary for experimentation
- materials necessary for display (posterboard, markers, etc.)
- books (using the library can help avoid these costs)

Less Is More Judges often take creative use of materials into account when judging a science fair, and expensive projects probably will not be recognized as better just because they employ fancier equipment. The library and Internet provide resources with suggestions on how to make equipment using materials that are inexpensive or can be found around the house. Also, many businesses and institutions will lend equipment to young scientists. Encourage your child to write to local businesses for help in obtaining materials if the schedule allows enough time for him or her to receive a response. You may wish to consult a scientific supply company, such as those listed in the box below, in order to determine how much materials will cost.

> **TIP**
>
> The following is a list of a few scientific supply companies that specialize in laboratory equipment. These companies require that biological and chemical supplies be purchased by your child's school and shipped to his or her teacher.
>
> Carolina Biological Supply Co.
> 2600 York Road
> Burlington, NC 27215
> (800) 334-5551
>
> Science Kit and Boreal Labs
> 777 East Park Drive
> Tonawanda, NY 14150
> (800) 828-7777
>
> WARD's Natural Science Establishment, Inc.
> 5100 West Henrietta Road
> Rochester, NY 14585
> (800) 962-2660

✓ If your child is involved in a group project, try to monitor his or her individual participation in the project. Consult the timeline, and check your child's progress at the end of Phase 2.

Phase 3—Data Collection and Analysis: Getting Involved

> **DURING THIS PHASE, STUDENTS WILL**
>
> 1. conduct experiments safely
> 2. choose an appropriate sample size
> 3. perform several trials of the experiment
> 4. collect data accurately
> 5. record all data and observations in their science project journal
> 6. graph or chart the data and look for trends
> 7. prepare a written conclusion supported by the data

Although your child's science teacher will plan for some in-class data collection, much of the data collection phase of the project may be done at home. Your child has been told that an adult must be present during all data collection. For your child's safety, please supervise the data collection phase. You may refer to the Safety Guide in the student resource packet to help avoid accidents during data collection. This is a good time to review the Safety Contract with your child.

Over the next few weeks, you may want to help your child in the following ways:

- Familiarize yourself with the Phase 3 student instruction sheets, and help your child understand the importance of each technique, such as using an adequate sample size, collecting the right type of data, and performing several trials of an experiment.

- Make sure that your child is measuring accurately.

- The International System of Units (SI) is a global measurement system that helps scientists share and compare their observations and results. If there is a requirement that your child express data measurements using SI, make sure that he or she understands this requirement.

- Check your child's science project journal to make sure that he or she is keeping a neat record of all aspects of the project.

- Assist in explaining charts, graphs, or basic concepts of analysis when necessary.

- Motivate your child by showing interest and asking questions, and allow him or her to decide how to conduct the experiment.

✓ If your child is involved in a group project, try to monitor his or her individual participation in the project. Consult the timeline, and check your child's progress at the end of Phase 3.

SCIENCE FAIR GUIDE **71**

Phase 4—Writing a Report: Getting Involved

> **DURING THIS PHASE, STUDENTS WILL**
>
> 1. answer the questions on page 55 of Resources for Students
> 2. prepare an outline and discuss it with the teacher
> 3. prepare a draft and discuss it with the teacher
> 4. revise the draft according to the teacher's feedback
> 5. turn in a completed draft

When your child's teacher assigns the written report, you may want to be sure that your child understands what is expected. He or she needs to follow the rules for how long the report should be, whether it must be typed, etc., especially if a written report is required for the science fair. You may want to contact the science teacher for report guidelines. The written report will represent your child's ideas and conclusions about the project, so it should be well thought out and neat.

The Process

Students will be asked to answer the questions on page 55 of Resources for Students. The questions are designed to get students thinking about what they learned and what they enjoyed the most about the science project. After answering the questions, students will create an outline of the information to include in their written reports. Using index cards may help with the organization part of the assignment. Once students have completed their outline, the teacher will meet with them to offer constructive criticism. From there, students will create at least one rough draft.

Positive Feedback If you help your child with an outline or draft of his or her report, remember that giving and receiving feedback can be difficult for parent and child. Here are some tips to aid the process:

- You may want to mention both the strong and weak points of an outline or a draft.

- You can certainly help your child with neatness, spelling, and grammar, but the ideas and writing should be his or her own.

- Remind your child to cite all sources he or she used. In the bibliography, your child should also acknowledge the help he or she received from parents, teachers, and other people.

- Be aware that your child's writing ability can affect the overall presentation of information. Imperfections in writing style may overshadow the scientific content. You may wish to check your child's report to ensure that it is factual and organized.

What to Include in the Written Report

Most scientific reports follow the same order as the steps of the scientific method, explaining the entire experimental process from beginning to end. Student reports should be ordered as follows:

- Your child may include some background information about the topic before stating the purpose and hypothesis of the project.

- A description of the procedure comes next, followed by the data acquired during the experimentation.

- Your child may choose to include charts, graphs, or photographs, but it is best to save most figures for the display.

- Last, your child will describe the conclusions that resulted from the data analysis. The conclusion is also the place to address some of the questions asked on page 55 of Resources for Students.

- You might also encourage your child to include suggestions for further study or a brief description of how he or she would do the project differently next time.

- Invite your child to have fun with the written part of the project.

✓ If your child is involved in a group project, try to monitor his or her individual participation in the project. Consult the timeline, and check your child's progress at the end of Phase 4.

Phase 5—Creating and Exhibiting a Display: Getting Involved

> **DURING THIS PHASE, STUDENTS WILL**
>
> 1. sketch possible designs for the display
> 2. create a display board within the appropriate parameters
> 3. display results in a clear and interesting manner
> 4. give an oral presentation as practice for the science fair interview

The Display Board

The purpose of a display is to present the information in the clearest manner possible so that the judges will be able to recognize quickly that the student performed a successful science project. You may want to encourage your child to be creative with borders, fonts, and layout, but have them make sure that a judge would find it easy to read the information contained in the display. Illustrations should be informative, not just decorative. A sleek, mature, and professional style can impress the judges, but they prefer presentations which are unique and clearly a product of the student.

Portable A display board can have a large middle section and two smaller "wings" on the left and right. This design allows the display to fold up, making it more portable. If the science fair is located far away, you might ask other parents to share the expense of renting a moving truck or trailer in order to save many families the worry of transporting displays that don't fit comfortably into cars.

> **HOW CAN I HELP?**
>
> - Encourage your child to sketch out a design before he or she begins making the display.
> - Check charts and graphs for accuracy and readability.
> - Make sure that the information shown is relevant to the purpose, hypothesis, and conclusion.
> - Remember that judges are looking for conclusions based on experimental evidence, not on opinions.
> - Give your child a pep talk before the interview.
> - Tell your child some tricks that you use to remain calm and focused.

Preparing for the Interview

Find out if your child needs to make a presentation to the judges at the science fair. The presentation summarizes each step of the science project: why students chose their subject, a statement of the hypothesis, what type of data was collected, a brief summary of the data, and the conclusions that students came to when they analyzed the data. Your child may also discuss how he or she would do the experiment differently if he or she were to start over again or what other questions arose during his or her research.

Practice Makes Perfect If your child practices a few times he or she will have a great advantage over students who haven't rehearsed. Your child could practice by explaining the project to your family and friends. During the interview, the judges will likely ask the student a few questions. If you ask a few questions during your child's practice presentations, he or she may feel more comfortable answering the judge's questions.

✓ If your child is involved in a group project, try to monitor his or her individual participation in the project. Consult the timeline, and check your child's progress at the end of Phase 5.

Be Supportive

Hopefully, every student will be recognized for his or her effort at the science fair. To show your support for your child, try to attend the science fair if possible and be proud of his or her individual effort and the part you played in his or her growth as a student.